Fire Your Film Crew and Use AI

How to Use AI to Write, Direct, Film, Design, Cast and Market your Films

Jim Bock

Contents

SECTION 1: THE FOUNDATION......3
Chapter 1: The Morning Everything Changed......4
Chapter 2: Your New Crew......14
Chapter 3: The AI Film Lab That Fits on a Desk......31
Chapter 4: The Director and the Technician......41
Chapter 5: The Logline and the Leap......48
Chapter 6: Characters, Dialogue, and the AI Writer's Room......54
Chapter 7: The Lock — and the Freedom It Gives You......61
Chapter 8: Seeing the Film Before It Exists......71
SECTION 2: PRODUCTION......80
Chapter 9: What Project Two Feels Like......81
Chapter 10: Scaling Up — Series, Anthologies, and Features......89
Chapter 11: Where This Is All Heading......97
Chapter 12: The 90-Day AI Film Plan......104
SECTION 3: RIGHTS, AUDIENCE, AND SELLING YOUR FILM......116
Chapter 13: The Trailer as Seduction......117
Chapter 14: The Audience You Build Before the Film Is Finished......125
Chapter 15: Festivals, Screens, and the Smell of Wet Popcorn......134
Chapter 16: Distribution — Getting Your Film Off Your Hard Drive...143
Chapter 17: AI Dubbing — The World Speaks Your Language......153
Chapter 18: The Business That Finds You......157
Chapter 19: Branded Content and the Generative Gold Mine......164
Chapter 20: Selling Without Selling Your Soul......168
Chapter 21: The Legal Stuff Without the Panic......176
Chapter 22: OpenClaw, Hermes, and the Rebel's Toolkit......185
Chapter 23: The Stigma — and Why It Doesn't Matter......190

SECTION 1: THE FOUNDATION

Writing, Planning, and Knowing Your Tools

The reader starts here knowing nothing except that they want to make films with AI. By the end of Section 1, they have a locked screenplay, a visual plan, and a complete understanding of what every tool in their lab does.

My name is Jim Bock. I'm sixty-five years old, and I've been making films for twenty-eight years. I started in traditional production — crews, cameras, locations, the whole expensive ecosystem — and I migrated to AI filmmaking when the technology finally caught up with the stories I wanted to tell. I run NJ Film Productions from my house in New Jersey, which means I make films in the same room where I eat breakfast. This book is everything I've learned about making professional-quality films with AI tools, written for anyone who wants to tell stories without waiting for permission.

Chapter 1: The Morning Everything Changed

Let me tell you about the moment I knew this was real.

It was a Tuesday morning in March. I was sitting at my kitchen table in New Jersey, drinking coffee that had gone lukewarm because I'd forgotten about it — something that happens more often at sixty-five than I care to admit — I've been making films for twenty-eight years, after all, and some things don't get easier. My wife had left to go to the gym. The dishwasher was running. And on my laptop screen, a man who looked exactly like me was delivering a monologue I'd written the night before, in a voice that sounded exactly like mine, in a room I'd described with words because I'd never been there and never would be.

I watched this digital version of myself blink, breathe, gesture with his hands the way I do when I'm making a point. The lip sync was tight. The lighting looked like someone had spent an hour placing lights. And I had made this without leaving my chair. No crew. No camera. No location scout. No call sheet. No fifty-thousand-dollar budget and no praying that the weather would hold.

I sat there for a long minute, staring at the screen. Then I finished my cold coffee and thought: Well, hell. The whole game just changed.

That was the morning I understood that AI filmmaking is not a trick. It's not a gimmick for TikTok kids. It is a genuine industrial shift — the kind that comes along once in a generation, like when sound hit silent movies or when digital replaced film stock. If you're my age, you've seen a few of these shifts. If you're younger, this might be your first one. Either way, I'm here to tell you: this is real, it works right now, and a single person at a desk can do what used to require forty people, five trucks, and a budget that would buy a house.

But let's not get ahead of ourselves. Before you start imagining yourself as the next Spielberg with nothing but a laptop and a dream, you need to understand what this technology actually is, what it actually does, and — just as important — what it cannot do no matter how many people try to sell you otherwise.

The Digital Twin: Your Stand-In That Never Tires

The heart of AI filmmaking — at least the way I do it at NJ Film Productions — is something called a Digital Twin. Think of it like this. In old Hollywood, every major star had a stand-in. While the crew lit the set and positioned the camera, the stand-in would sit in the actor's chair so the real actor could rest in their trailer, read a magazine, have a sandwich. The stand-in looked enough like the star to be useful for technical setup, but they weren't the star. They were a placeholder.

A Digital Twin is like that stand-in, except it looks exactly like you — or like whoever you train it on — and it can perform. Forever. Without getting tired, without aging, without showing up late because traffic was bad on the 405. You record a fifteen-second video of yourself talking naturally, upload it to a tool called HeyGen along with a clear photograph, and the system learns what you look like and how you move. Twenty minutes later, you have a digital copy of yourself that can say anything you type into a script box.

Grumpy Old Man Tool Note: HeyGen

HeyGen is useful because it turns the impossible part of small filmmaking - getting a consistent person on camera whenever you need them - into a repeatable workflow. In the old world, you needed a camera, lights, sound, a room, a shirt that was not wrinkled, and a human being willing to say the same line six different ways. Now you can build an avatar, feed it a script, and get usable dialogue scenes while sitting at the same desk where you pay the electric bill. That is not a gimmick. That is production leverage.

The possibility here is bigger than a talking head. You can create a recurring host, a narrator, a fake news anchor inside your story, a company training character, a documentary guide, a professor, a detective, a therapist, a villain doing a confession, or a version of yourself explaining the film before the film exists. Pair it with ElevenLabs for stronger voice performance, Seedance for atmosphere, and DaVinci Resolve for editing, and suddenly you are not making little avatar clips. You are building a repeatable scene machine.

The limitation is movement. Do not ask HeyGen to do gymnastics, sword fights, car chases, or emotional acting that requires the body language of a trained stage actor. That is like asking a toaster to mow the lawn. Use HeyGen for faces, dialogue, consistency, and controlled performance. Design around the strength of the tool and it will make you look smart. Ignore the limitation and the footage will look like a mannequin trying to win an Oscar.

I want you to sit with that for a second. Anything you type. Any script. Any language. Any emotional register. Your Digital Twin will say it, with your face, your expressions, your mannerisms, in a location you describe with words. A Parisian apartment at golden hour? A stark interrogation room? A spaceship corridor? You describe it, and your Twin appears there, saying your words, wearing whatever outfit you specify.

The technology that makes this possible is called a neural network, which is just a fancy way of saying "a computer system that learned by looking at millions of examples." It studied countless videos of human faces talking, learned the relationship between sounds and mouth shapes, between words and facial expressions, between emotional tone and body language. Then you give it your specific face and voice, and it applies everything it learned to make your face move in a way that looks alive.

This matters because the face is everything in film. We are wired to read faces. A flicker of an eyebrow, the timing of a blink, the way someone's mouth tightens before they say something hard — this is where cinema lives. The Digital Twin doesn't replace a human actor for every kind of performance. Complex physical acting, fight choreography, subtle emotional arcs that require dozens of micro-expressions — that's still better handled by real people on real sets. But for dialogue? For teaching? For testimony, confession, argument, persuasion, the quiet scene where someone tells the truth they've been hiding? The Digital Twin is already good enough. With proper direction and voice work, it's better than good. It's convincing.

Your Voice, Working the Night Shift

Now here's where it gets spooky in the best way. The face is only half the performance. The other half is the voice, and voice cloning technology — specifically a tool called ElevenLabs — has crossed a threshold that I honestly didn't think we'd reach this soon.

You record a few minutes of yourself talking naturally. Not reading stiffly off a page — just talking, the way you'd explain something to a friend at the diner. ElevenLabs listens to that recording, maps the unique shape of your vocal cords, your speech patterns, your rhythm, the way you emphasize certain words, even the little pauses where you breathe. And then it can generate new speech in your voice. Any words. Any emotion. You can whisper, shout, speak with grief or joy or quiet menace, all in a voice that is recognizably, unmistakably yours.

I remember the first time I heard my cloned voice say something I had never actually said out loud. It was a short test monologue about grief, memory, and a man sitting alone in a home studio at three in the morning. I typed the line into ElevenLabs, added an audio tag to make it sound whispered and broken, hit generate, and heard my own voice deliver the line with emotional precision that surprised me. It gave me chills. And I don't spook easy. I'm sixty-five. I've seen things.

The practical implication is profound. Your voice becomes an asset that works while you sleep. You can generate hours of dialogue, iterate on performance choices, try a line ten different ways and pick the one that lands. In traditional production, every take costs money — the actor's time, the studio rental, the engineer's hourly rate. In AI production, iteration is nearly free. You adjust a few words in the prompt and generate again. The cost of trying something different drops to almost nothing, which means you can experiment your way to better work instead of committing to your first idea because you can't afford a second take.

Grumpy Old Man Tool Note: ElevenLabs

ElevenLabs is where the voice starts working the night shift. That matters because audiences forgive a lot of visual weirdness, but they do not forgive bad sound. Bad sound makes even a good scene feel like it was recorded inside a soup can. With ElevenLabs, you can build narration, character voices, scratch dialogue, trailer voiceover, temp performances, alternate line readings, and foreign-language tests without booking a studio or bribing your nephew to read lines in the garage.

The real possibility is performance iteration. Try the line angry. Try it tired. Try it whispered. Try it like a man admitting something he should have admitted ten years ago. Then take the best voice pass into HeyGen, cut it against Seedance B-roll, and shape the whole thing in Resolve. That is how a small filmmaker starts acting like a department instead of a lonely guy yelling at a laptop.

The warning is consent and taste. Do not clone somebody's voice without permission unless you enjoy legal letters written by people with expensive pens. And do not assume the first generated take is the final take. Sometimes the voice sounds polished but emotionally dead, like a bank manager reading a sympathy card. Direct the voice. Regenerate. Compare. Use your ears. They are still part of the crew.

The Camera That Exists in Software

So you have your actor — the Digital Twin. You have your voice — the clone. Now you need the world around them. The room, the light, the atmosphere, the B-roll that makes a film feel cinematic rather than just functional.

This is where text-to-video generation comes in, and this is the part that still feels like science fiction even though I've been using it for two years. You type a description — "cinematic wide shot of a cluttered workroom at night, warm amber light from a single desk lamp, audio equipment stacked on metal shelves, shadows deep and textured" — and a tool called Seedance generates a moving image that matches your description. Not a still photograph. A video. Five, ten, twenty seconds of footage with camera movement, atmospheric detail, light that behaves like real light.

Think about what this means. In traditional filmmaking, that shot would require a location scout to find the room, a production designer to dress it, a gaffer to light it, a cinematographer to frame it, a camera operator to execute the move, and a full crew to support all of the above. Days of preparation. Thousands of dollars. In AI filmmaking, you describe what you want and the camera — which is just software running on a server somewhere — builds it for you in a few minutes.

The quality varies. Some generations come out looking like something from a thirty-million-dollar feature. Others come out looking like a video game cutscene from 2019. The difference is almost always in the prompt — the specificity and clarity of your description — and in your taste, your ability to look at ten generated options and know which one serves the story. But the ceiling is astonishingly high, and it rises every few months as these companies release new models.

Grumpy Old Man Tool Note: Seedance

Seedance is one of the tools that makes AI filmmaking feel like somebody stole the keys to the studio truck and handed them to regular people. It can give you atmosphere, camera movement, establishing shots, street corners, lonely rooms, rainy windows, dramatic inserts, and the kind of visual glue that makes a film feel like a film instead of a slide show with ambition.

Think bigger than pretty B-roll. Seedance can become your location scout, second-unit camera, mood machine, dream-sequence engine, trailer footage generator, and world-building department. Use Midjourney first to define the look, then Seedance to make the world move. Add ElevenLabs dialogue, HeyGen performance, and Resolve editing, and you have a pipeline that can make a one-person project feel like it had people with walkie-talkies standing around looking serious.

The limitation is control. AI video will occasionally give you a shot that looks expensive and then ruin one hand, one doorway, or one piece of physics just enough to make you mutter at the screen. Welcome to filmmaking. Traditional crews ruin shots too; they just do it with catering. The answer is not panic. Generate variations, cut around the weak spots, and never let one bad frame boss you around.

Pipeline A and Pipeline B: The Two Ways to Build a Scene

Now we need to talk about workflow, because this is where most beginners get tangled up. There are two basic pipelines for putting these tools together, and understanding which one to use when is the difference between a project that flows and a project that stalls.

Pipeline A is what I use for about eighty percent of my work. You generate each character separately in HeyGen — close-up shots, medium shots, reaction shots — and then you cut them together in editing software the way you would cut together footage from a traditional shoot. Character A says their line in a close-up. Cut to Character B's reaction. Cut to Character A listening. Cut to a wide shot generated in Seedance that shows both of them in the room together. This approach gives you the highest quality per dollar spent, the most control over performance, and the most forgiving workflow for beginners. It is how I made the sample short, the twelve-minute thriller that proved this whole system works.

A dialogue-driven sample short is the kind of project this workflow was built for: three characters, mostly one location, heavy on dialogue and sound design. That is where AI avatars shine. The point is not to pretend the computer can do everything. The point is to design the film around what the tools do well: faces, voices, atmosphere, close-ups, emotional monologues, and controlled interiors.

Pipeline B is for those moments when two or three characters genuinely need to be in the same frame — sharing eye contact, reacting to each other in real time, occupying the same physical space. HeyGen has a feature called Avatar Shots that uses Seedance as the underlying engine to composite multiple avatars into a single shared scene. It costs more in credits, and there are time limits to work within, but the result looks cinematic in a way that separate generations cut together cannot quite match.

The smart project uses both. Pipeline B for one or two key shared-frame moments per video — the confrontation, the confession, the moment where two people look at each other and everything changes. Pipeline A everywhere else. This is how you manage budget and quality together.

The Math That Made Me a Believer

Let me talk numbers for a minute, because this is where the skeptic in me — and I was a skeptic, make no mistake — finally had to admit that something fundamental had shifted.

A traditional twelve-minute dialogue short can easily price out at tens of thousands of dollars once you add a skeleton crew, actors, locations, sound, lighting, insurance, food, editing, color, and festival delivery. Even the "let's beg everyone for favors" version can become expensive fast. That is why so many good short films never get made. They die in the drawer, not because the story is bad, but because the math is rotten.

With AI, that same kind of short can be built for a fraction of the traditional cost. Generated environments replace location rentals. Digital Twins replace some performance needs. Voice cloning reduces studio time. Music, B-roll, and visual concepts can be produced in hours instead of weeks. You still need taste, judgment, writing, editing, and patience. But the money wall is lower than it has ever been.

The difference is not marginal. It is not "slightly more accessible." It is the difference between buying a house and buying a nice dinner. And that difference changes everything about who gets to make films and what kinds of films get made.

Think about it this way. Traditional production has a minimum viable budget beneath which the project literally cannot function. You cannot shoot a film with zero lights, zero crew, and zero equipment unless you already own professional gear and know people willing to work for free. There is a floor, and that floor is high — high enough to exclude most people who have stories to tell but don't have ten thousand dollars sitting in their checking account.

AI production has no fixed floor. You can spend fifty dollars and make something. It may not be good, but it can exist. And fifty dollars of something that exists is infinitely more valuable than fifty thousand dollars of something that lives in a drawer forever.

What AI Removes — And What It Leaves Standing

I want to be straight with you about what is real here and what is hype, because I have seen too many beginners crash on the rocks of unreasonable expectations.

AI removes logistics. You do not need to coordinate eight people's schedules. You do not need to rent a truck. You do not need to beg a restaurant owner to let you shoot during their closed hours. The work happens at your desk, on your timeline, at your pace. If you're like me — sixty-five years old, bad back, not interested in hauling equipment up stairs — this is not a minor convenience. It is the reason you can keep making work at all.

AI removes physical constraints. You can put a scene in a Parisian apartment even if you live in Ohio and have never been to Paris. You can have a character wear a bespoke suit that would cost thousands to own. You can stage a conversation on a rooftop overlooking a cityscape that would require permits, safety harnesses, and liability insurance to capture with a real camera. The world becomes a prompt away.

AI removes iteration cost. In traditional production, reshooting a scene means reconvening the crew, rebooking the location, hoping the actor's haircut hasn't changed, praying for the same weather. In AI production, you adjust a prompt and regenerate. If the lighting in shot eleven doesn't match shot three, you don't schedule a reshoot day. You rerender. The cost of fixing a mistake drops to nearly zero, which means you can afford to be a perfectionist in ways that traditional production would bankrupt.

AI removes the gatekeeping. For decades, filmmaking has been a club with a steep membership fee. You needed access to capital, equipment, locations, and people with specialized skills. AI doesn't make the club disappear — there will always be room for big productions with real crews making spectacular work — but it opens a side door. A big one. And through that door walks everyone who has a story and the discipline to finish it.

But here is what AI cannot do, and I want you to read this carefully because this is where people get themselves in trouble.

AI cannot give you taste. Taste is the ability to look at ten generated images and know which one serves the story. Taste is knowing when a line of dialogue is fake even if it's grammatically perfect. Taste is cutting the scene you love because it slows the film down. No tool provides this. You develop it by making work, watching work, and being honest about what landed and what didn't. Taste is earned, not purchased, and certainly not generated.

AI cannot give you something to say. The best AI-generated film with no point of view is still a film with no point of view. A beautiful image that means nothing is just a screensaver. You must bring the reason for the film to exist — the question it asks, the emotion it explores, the perspective you offer that the viewer hasn't seen before. The tools handle execution. You handle intention. That division of labor is non-negotiable.

AI cannot give you judgment. Judgment is the willingness to kill your darlings, to admit the second act is sagging, to recognize that the voice you cloned sounds slightly off in one register and find a workaround. Judgment is knowing when the tool has given you something adequate and pushing it to something good. Judgment is finishing the project instead of endlessly regenerating in search of perfection that does not exist.

AI cannot give you storytelling instinct. Storytelling instinct is the felt sense of when to cut, when to hold, when to silence the music, when to let a character breathe. It is knowing that the audience is ahead of you and needing to surprise them, or that the audience is lost and needing a handhold. This instinct comes from consuming stories critically and making stories repeatedly. No prompt engineering substitutes for it.

The Possibilities That Nobody Is Talking About

Now let me open some doors you might not have realized were there, because this is the part that excites me — the grumpy old man who is secretly optimistic about all of this.

Heavy dialogue films are the obvious sweet spot. Stage adaptations. Courtroom dramas. Two people in a room having the conversation that changes everything. These formats play directly to the strengths of avatar-based production. But think bigger. Think about documentary-style interviews where the subject is no longer available — historical figures brought to life with respectful, consented recreations. Think about educational content where a professor can generate a semester's worth of lectures in a week, in forty languages, without recording forty separate versions. Think about training videos for corporations that currently spend fifty thousand dollars on a single safety video that nobody watches.

Think about branded content, anthology series, podcast-to-video conversion, therapy resources, language learning, testimony preservation — the list goes on and on.

Here's one that keeps me up at night in a good way: what if you used AI filmmaking for scenes that are impossible to shoot traditionally? A character hallucinating, and the world around them slowly dissolving into something else — Pika's stylized transformation effects layered over Seedance B-roll, creating a visual experience that would cost a VFX house a hundred thousand dollars to produce. A dream sequence where Luma's keyframe interpolation morphs one location into another seamlessly. A title sequence generated with Pika's distinctive aesthetic that establishes the film's visual language before the first frame of story. These aren't just possible — they're waiting for someone with imagination to execute them.

But I need to be honest about the limitations, too. Avatars are not for action movies with stunts. They're not for films where complex physical choreography carries the story. A fight scene between two avatars will look, at best, like a video game cutscene and, at worst, like a physics experiment gone wrong. If your story requires sustained full-body athleticism, hire real actors and a stunt coordinator. The technology will get there eventually — probably sooner than I think, given how fast everything else has moved — but it's not there yet.

The person who wins in this new landscape is not the person who discovers the most powerful AI tool first. Tools change monthly. The person who wins is the one who knows what they want, can describe it clearly, can judge the result honestly, and can finish the project instead of endlessly experimenting. That sentence is the operating system of this entire book. Read it again. Everything else is implementation detail.

What Comes Next

In the next chapter, I'm going to walk you through every tool in the AI filmmaker's toolkit — what each one does, where it fits in the pipeline, what it's best at, and when to use it instead of something else. We're talking about ChatGPT and Claude as your writers, Midjourney as your concept artist, HeyGen as your casting director and lead actor, ElevenLabs as your voice department, Seedance as your camera crew, and a whole constellation of supporting tools that handle music, sound, editing, color, and finishing.

But before you turn that page, I want you to sit with something. For your entire life, if you wanted to make a film, you needed permission. Permission in the form of money, access, connections, equipment, locations, people's time. That permission structure just got dismantled. Not entirely — making great work still requires skill, taste, discipline, and something to say. But the barrier that kept most people from ever trying, from ever finishing, from ever showing their work to an audience? That barrier is gone.

What are you going to make now that you don't need permission anymore?

That's the real question this book answers. The tools are just tools. The filmmaking is up to you.

Chapter 2: Your New Crew

When I made my first short film in 2018, I hired a three-person crew for two days. The cinematographer was eight hundred a day plus gear rental. The sound mixer was six-fifty. The PA got a hundred and fifty to handle coffee runs. Fourteen hours of labor, forty-three minutes of usable footage, and a four- thousand-dollar hole in my checking account. That project never got finished. The hard drives gathered dust in my desk drawer next to my reading glasses and coupons I keep forgetting to use. These days my crew is a stack of software subscriptions. No overtime. No sick days. No parking. And the films actually get finished, because the biggest barrier was never creativity — it was logistics. Remove the logistics and the work flows.

This chapter is your orientation. I'm walking you through every tool in the AI filmmaker's toolkit — what each one does, where it fits, and when to use it instead of something else. Think of this as meeting your new crew on day one of production.

Your Writers: ChatGPT and Claude

Every film starts with words. Before you generate a single image, before you open a video tool, before you spend your first rendering credit, you need a story. ChatGPT and Claude — two AI language models that you access through a web browser — are your writers' room. I use ChatGPT for brainstorming and structure. It's fast and prolific — twenty logline variations in thirty seconds, a beat sheet that doesn't sag, dialogue that sounds like humans actually talk. When I'm stuck on a story problem, ChatGPT generates alternatives until something clicks.

Claude — made by Anthropic — is my go-to for longer work. It has a bigger memory, which means it can hold an entire feature screenplay and talk about it coherently.

When subtext matters — where what's not said is as important as what is — Claude gives me the more psychologically layered version. ChatGPT for speed and volume. Claude for depth and nuance. I use both. These tools are not writing your film for you. They're the colleagues who throw out ideas while you make every decision. But having a writers' room that never gets tired and can generate twenty versions while you refill your mug? That changes the game. Here's a possibility most people miss: use these tools to generate the emotional direction for scenes, not just dialogue. Feed it your script and ask: "What should the audience feel here? What sound design supports that?

What color temperature should the lighting have?" These models have read more film theory than any human alive. Use that knowledge.

Grumpy Old Man Tool Note: Claude

Claude is the calmer reader in the corner, the one who actually remembers what happened in Act One. That makes it valuable for longer scripts, treatments, book chapters, show bibles, and complicated story problems. ChatGPT is often better when you need speed. Claude is often better when you need the whole thing held in one head without dropping pieces on the floor like a toddler with groceries.

Use Claude when you want structure, emotional logic, continuity, and honest diagnosis. Ask it whether the character arc earns the ending. Ask where the middle sags. Ask which scene is pretending to be important but is really just furniture. Then compare its advice against ChatGPT or your own notes. The future is not one model replacing your brain. The future is you running a small advisory board from a laptop.

Claude can still be too polite. Sometimes it wraps criticism in a warm blanket when what you need is a slap on the hood and a mechanic saying, 'This thing is leaking oil.' So ask directly. Tell it to be blunt. Tell it to protect the audience, not your feelings. A good AI reader should help you fix the movie, not compliment the wreckage.

Grumpy Old Man Tool Note: ChatGPT

ChatGPT is the fast-talking assistant who can throw twenty ideas at the wall before you finish your coffee. That is useful, because most filmmakers do not get stuck because they have no talent. They get stuck because their idea is still fog. ChatGPT helps turn fog into options: loglines, beats, alternate endings, character motives, trailer hooks, pitch copy, and the dreaded plain-English answer to, 'What is this movie actually about?'

The possibility is using ChatGPT as a creative pressure washer. Ask it to give you five cheaper ways to shoot the scene, ten stronger openings, three endings that do not feel like oatmeal, or a production plan for a short film built entirely around one room and two voices. Then feed those options to Claude for deeper critique, to Midjourney for visual direction, or to Seedance for shot experiments. One tool is useful. A conversation between tools is where things get interesting.

Do not mistake speed for genius. ChatGPT will sometimes hand you dialogue so clean and lifeless it sounds like a corporate training video on a cruise ship. Push back. Ask for rougher, stranger, more human, more specific. It is not your screenwriter. It is your tireless junior assistant. You still have to be the cranky adult in the room.

Your Concept Artist: Midjourney

Once you have your story, you need to see it. Midjourney takes text descriptions and produces stunning visual interpretations. I use it for three critical things: character reference portraits, location packs, and storyboard panels. A location pack is the same setting from multiple angles — wide, medium, close-up details — rendered in a consistent style. You generate these before you touch a video tool, and they become your visual anchor. Without this step, your film looks like a Pinterest board instead of a movie. The quality is outrageous. I've generated images I'd have paid a concept artist five hundred dollars apiece to produce.

The trick is in the prompt — the specificity of your description — and in your taste, your ability to look at ten variations and know which one serves the story. Here's what excites me: Midjourney is also your look development department. Before you commit to a visual style, generate twenty different lighting approaches and see what each does to the emotional temperature. A hospital room feels radically different under warm golden-hour light versus cold fluorescent versus the amber of a single bedside lamp. You can preview all of them before generating a single frame of video. That used to be the domain of big-budget previs departments. Now it's a prompt away.

Grumpy Old Man Tool Note: Midjourney

Midjourney is the concept artist who does not need a studio, a retainer, or three weeks to 'circle back.' You type a visual idea and it gives you something you can look at. That is a bigger deal than beginners realize, because film is a visual medium and most people cannot judge what you mean until they see it. A blurry idea in your head is not a production design. It is a weather condition.

Use Midjourney to build mood boards, character portraits, location references, poster concepts, storyboard inspiration, wardrobe ideas, lighting styles, and the overall visual language of the film. Then feed that look into Seedance, Runway, or Sora as reference thinking. It becomes the visual constitution of the project. Everybody - including the AI tools - needs to know what country they are living in.

The danger is that Midjourney makes everything look too pretty. Pretty can be poison. A gritty recovery drama does not need to look like a perfume ad for vampires. Pick images that serve the story, not the ones that make you feel clever. Taste is the filter. Without taste, Midjourney is just an expensive slot machine with nice lighting.

Your Casting Director and Lead Actor: HeyGen

HeyGen is the platform where your Digital Twin lives, performs, and comes to life. I covered the concept of Digital Twins in the last chapter, but now let's talk about what HeyGen actually does in production terms. Think of HeyGen as your casting director, your lead actor, and your on-set playback system rolled into one. You create your Digital Twin — or multiple Twins for different characters — and then you write scripts for them. The Twin performs the script, lip-syncing to whatever voice you've assigned, in whatever Look you've designed. You can change their outfit, their background, their camera angle, all without re-recording anything. HeyGen's Photo Avatar feature deserves special mention.

If you have a still photograph of a person — with proper permission and consent documentation — you can animate it. The person in the photo will speak your script, with natural head movement and expressions. It's not as convincing as a full Digital Twin trained on video, but for supporting characters who only appear briefly, or for historical recreations using public-domain photographs, it's incredibly useful. And here's a possibility spark that keeps me up at night: what about using Photo Avatars for the "unfilmable" characters? The ghost in a memory play. The person on the other end of a video call. The photograph that comes to life in a fantasy sequence. These are uses that play to the tool's strengths while acknowledging its limitations.

Smart design beats brute force every time.

Grumpy Old Man Tool Note: HeyGen in the Pipeline

In the larger pipeline, HeyGen is not just an avatar toy. It is casting, continuity, and performance control all rolled into one. Once you create a character, that character can return again and again. Same face, same general framing, same usefulness. Try getting that from a real actor after three reschedules, a haircut, and a sudden conflict with their cousin's wedding.

The future use is serialized storytelling. A recurring host. A detective who appears across episodes. A corporate spokesperson. A teacher. A fake documentary witness. A digital version of you explaining the process behind each film. Combine HeyGen with ElevenLabs voices, Seedance environments, and a proper editing workflow, and you can build a small universe instead of one lonely video.

The rule is simple: make the scene fit the tool. Close-ups, medium shots, testimony, confessions, arguments, instructions, and screen-based conversations work. Full-body ballet with falling debris does not. If you design intelligently, HeyGen saves you money. If you design like a man trying to tow a boat with a bicycle, do not blame the bicycle.

Your Voice Department: ElevenLabs

If HeyGen provides the face, ElevenLabs provides the voice, and I cannot stress enough how important the voice is. A mediocre face with a great voice still delivers a performance. A great face with a mediocre voice looks like a puppet. ElevenLabs offers two voice clones. The Instant Voice Clone takes about a minute of clean audio and produces a voice that's ninety percent accurate in minutes. This is your workhorse — fast, responsive to emotional direction through audio tags in your script. The Professional Voice Clone requires thirty minutes to several hours of training audio and produces broadcast-grade fidelity. For a feature where the voice carries the weight, the Professional clone is worth the setup.

The audio tag system is where ElevenLabs shines. You write tags in square brackets — [whispers], [sighs], [angry] — and the voice engine interprets them. I've adjusted a line's emotional calibration by changing a single word in a tag and heard the difference immediately. That used to require a recording booth and hours of takes. A warning: ElevenLabs charges by character count, not by minute. A feature-length screenplay will burn through characters fast. Check your dashboard before a big render. Nothing kills momentum like your voice stopping because you've hit your limit.

Grumpy Old Man Tool Note: ElevenLabs in the Pipeline

ElevenLabs becomes especially powerful when you stop thinking of it as 'text to speech' and start thinking of it as a voice department. Narrator, side character, inner monologue, phone message, radio broadcast, documentary interview, villain tape, ghostly memory - all of these can be built as audio first. In film, audio often tells the audience what the picture is too expensive to show.

Pair ElevenLabs with Descript for rough assembly, HeyGen for spoken performance, Suno or Udio for musical beds, and Resolve for final mix. This is where the one-person filmmaker starts stacking tools like a retired guy stacking coupons at the supermarket. One coupon is nice. Ten coupons is dinner.

The limitation is emotional sameness. If every line is generated with the same rhythm, your film will sound like a polite hostage note. Use pauses. Use tags. Rewrite the sentence for speech, not for reading. A voice line has to breathe. The machine can generate it, but you still have to make it sound like a person with blood pressure and regrets.

Your Cinematic Camera Crew: Seedance

Seedance is a video generation tool made by ByteDance — the same company behind TikTok — and it is, in my opinion, the most cinematic of the current generation of AI video tools. Where HeyGen excels at faces and dialogue, Seedance excels at atmosphere, camera movement, and the kind of B-roll footage that makes a film feel like a film rather than a video call. I reach for Seedance when I need establishing shots — the wide view of the city at dusk, the interior of the bar before the characters enter, the hallway that stretches into shadow. I reach for it when I need camera movement — a slow push-in, a gentle drift, a handheld feel that suggests documentary realism.

I reach for it when I need atmosphere — rain on windows, smoke in streetlights, dust particles catching a shaft of afternoon sun. Seedance also has a native audio generation feature that is genuinely impressive. You can generate video with synchronized ambient sound — footsteps, room tone, environmental audio — baked right in. This doesn't replace proper sound design, but it's a fantastic starting point and can save hours of work. Here's a combination that most people haven't tried: use Seedance's audio generation for the ambient world-building, then layer ElevenLabs dialogue on top in your edit. The result is a scene with real spatial depth — the character's voice clean and close, the environment alive and present around them.

It feels like a movie because it sounds like a movie.

Your Reliable Sedan: Sora

OpenAI's Sora made waves when the first demos dropped, and for good reason — the physics simulation is remarkable. Water flows correctly. Smoke disperses naturally. A person walking through snow leaves footprints that make sense. Sora understands physical causality better than most generators. I think of Sora as my reliable sedan. Not the flashiest, but predictable. I use it for wide establishing shots and scenes with complex physics — water, fire, smoke, cloth in wind. The trade-off is that it's less "cinematic" than Seedance, less atmosphere and mood. But for grounding your film in physical reality, it's invaluable. Use Sora for the wide shots that establish reality, Seedance for the atmospheric inserts that give your film emotional color.

Together they cover the full cinematic range.

Grumpy Old Man Tool Note: Sora

Sora is useful because it aims at coherent, realistic video generation - the kind of shot where physics, movement, and environment need to feel less like a cartoon and more like the world behaving itself. I call it the reliable sedan because sometimes you do not need fireworks. You need the thing to start, drive straight, and not embarrass you at the intersection.

The possibility is using Sora for grounding shots: weather, water, streets, landscapes, wide establishing images, practical movement, and scenes where realism matters more than style. Use Seedance when you want mood, Kling when you need more physical action, and Sora when you need the world to feel believable enough that the audience stops inspecting the machinery.

But do not worship it. Every model has blind spots. A shot can look astonishing and still be wrong for your film. The job is not to use the fanciest tool. The job is to choose the shot that helps the story land. A reliable sedan is wonderful. It is still not a substitute for knowing where you are going.

Your Editor's Safety Net: Runway

Runway occupies a unique place in my toolkit because it's not primarily a "generate from scratch" tool for me — it's a "fix and modify" tool. Runway's Gen-4.5 model can take existing footage and change it. Different camera angle of the same scene. Extend a clip that ended too soon. Add an element that wasn't there. Remove an element that shouldn't be there. This is your safety net. When you've generated a shot that's eighty percent perfect but the camera is slightly too high, or the lighting is a touch too warm, or you need the character to be looking a different direction — Runway can often fix it without requiring a full regeneration. Runway also has Aleph, a model that excels at stylized, artistic interpretations.

If your film needs a dream sequence, a memory flash, or a title sequence with a distinct visual identity, Aleph is where I start.

Grumpy Old Man Tool Note: Runway

Runway is the tool I think of when a shot is almost right and needs help. Almost right is where filmmakers live. The light is close, the motion is close, the length is close, the idea is close, and then one little thing ruins your afternoon. Runway can help extend, modify, stylize, repair, or rework footage so you do not have to throw away a usable idea because one piece coughed at the wrong time.

Use it with restraint. Runway can support transitions, fixes, alternate versions, stylized memories, dream flashes, and cleanup work after Seedance, Sora, Kling, or Pika gives you a promising but imperfect clip. This is the digital version of a good post-production technician saying, 'Relax, we can probably fix that.' Those people are worth their weight in hard drives.

The limitation is rabbit holes. Runway makes it easy to keep improving a shot until you have improved it into a coma. Do not spend two days fixing a five-second insert nobody will remember. Use the tool to solve production problems, not to avoid finishing the film. Perfectionism is just procrastination wearing nicer shoes.

Your Physics Professor: Kling

Kling, from Kuaishou, is what I reach for when things move in physically complex ways. Fight chore- ography. A character running through rain. Cloth rippling. Water splashing. Kling's understanding of fluid dynamics, fabric simulation, and impact response is the best in the business. If you're making an action film, Kling is essential. I've generated fight scenes that looked more physically coherent than some big-budget VFX. The trick is describing the physics explicitly, not just the action. "A man punches through a wooden board, splinters flying in slow motion, dust catching backlight" beats "man breaks board" because it tells the physics engine what to simulate.

Grumpy Old Man Tool Note: Kling

Kling earns its place when movement matters. Running, falling, fighting, rain, cloth, impact, bodies in motion - these are the things that expose weak video generation fast. A bad static shot is boring. A bad action shot looks like a medical condition. Kling is worth testing when the scene needs motion that carries weight.

The possibilities are obvious: short action beats, chase fragments, storm footage, fight inserts, kinetic trailer moments, physical transitions, and sequences where you need the world to move with force. Use it alongside Seedance for atmosphere, Runway for fixes, and Resolve for aggressive cutting. You do not need to generate a perfect thirty-second fight. You need three good shots and the nerve to cut fast.

The limitation is complexity. The more bodies, props, impacts, and camera moves you demand, the more likely the machine will start making decisions like a drunk stagehand. Keep prompts focused. Generate short. Inspect everything. If a hand turns into a baked potato, do not argue with reality. Regenerate or cut around it.

Your Effects Specialist: Pika

Pika Labs makes a video generation tool that I keep in my back pocket for specific jobs: stylized trans- formations, quick visual effects, and the fastest iteration cycle of any tool in my stack. If Seedance is my cinematographer and Kling is my stunt coordinator, Pika is my effects specialist — the person who can whip up something impossible-looking in ten minutes. The transformation effects are where Pika really shines. A person dissolving into particles. A room melting into a dreamscape. Objects morphing into other objects. These are effects that would cost a VFX house thousands of dollars and weeks of render time. Pika generates them in minutes. Here's a use case that most people haven't considered: title sequences.

Pika's stylized aesthetic — slightly painterly, very dynamic, always visually interesting — is perfect for opening titles that need to establish a film's visual identity. Generate a sequence of abstract transformations that evoke your film's themes, cut them together with your title cards, add music. You've got an opening that feels expensive and distinctive, generated in an afternoon.

Grumpy Old Man Tool Note: Pika

Pika is the effects specialist - the tool you call when reality needs to bend a little. Transformations, stylized shots, surreal moments, animated transitions, visual metaphors, title sequences, and social teaser clips are where Pika can earn its keep. It is not always the main camera. Sometimes it is the strange cousin who shows up with smoke bombs and somehow improves the party.

The imaginative use is theme. What if a character's mental state slowly distorts the world? What if memories enter the film as visual transformations? What if the title sequence tells the audience, before a word is spoken, that reality is going to be unstable? Pika can help you create a visual language that would have cost real money in the old days.

The danger is using effects because you can. That is how films start looking like a demo reel at a trade show. The effect should mean something. If the room melts, the story better be melting too. Otherwise you are just making digital soup.

Your Transition Artist: Luma

Luma's Ray model — particularly the 3.14 version — has a feature that I use constantly: image-to-video with keyframe interpolation. You give it a starting image and an ending image, and it generates the motion between them. This is magic for transitions. A close-up of your character's face morphing into the wide shot of the landscape they're staring at. An object in one scene transforming into an object in the next scene, creating a visual bridge that connects two moments in time. Dream sequences love this technique. So do memory plays, where the present dissolves into the past. Luma's interpolation feels organic and continuous in a way that a simple crossfade never does. The motion has intention. It feels directed.

Grumpy Old Man Tool Note: Luma

Luma is valuable when you need movement between ideas. A still image becoming a moving moment. A photograph turning into memory. A location shifting into another location. A character crossing from one emotional state into another without you explaining it like a man reading appliance instructions.

This makes Luma useful for transitions, memory sequences, documentaries, dream passages, time jumps, and visual bridges between scenes. Pair Midjourney reference images with Luma motion, then bring the results into Resolve. Suddenly a cut is not just a cut. It becomes a piece of storytelling.

The limitation is subtle control. Luma may give you a beautiful motion that goes slightly the wrong way, like a GPS that insists your driveway is a highway. Use it when the motion supports the story. If the transition calls attention to itself more than the scene, trim it, simplify it, or throw it into the drawer marked 'nice but useless.' Every filmmaker needs that drawer.

The Supporting Cast: Veo, Frameo, LTX Studio, ImagineArt

Google's Veo is worth knowing for one feature: native synchronized audio generation. Its ambient sound feels more spatially precise — a real acoustic environment around the camera. I use it for shots where the sound of the space matters as much as the look. Frameo, LTX Studio, and ImagineArt are pipeline orchestrators — tools that handle multiple production stages in one interface. Script to storyboard to shot list to generation. For beginners who want the simplest workflow, an orchestrator gets you to a finished film faster. The trade-off is flexibility. Individual tools give you more options. Orchestrators sacrifice control for convenience.

Grumpy Old Man Tool Note: Veo, Frameo, LTX Studio, and ImagineArt

Veo, Frameo, LTX Studio, and ImagineArt sit in the broader supporting-tool category. They are the tools you test when your main pipeline needs something specific: synchronized sound, easier storyboarding, all-in-one workflow support, quick image generation, or a different flavor of video output. Not every project needs every tool. That is how subscription creep sneaks into your wallet wearing a friendly hat.

The possibility is workflow diversity. One tool might be better for ambient sound. Another might be better for organizing a script into shots. Another might help a beginner move from idea to storyboard without getting lost in seventeen browser tabs. These supporting tools can become bridges between writing, visual planning, generation, and editing. That matters because the future belongs to the filmmaker who can connect systems, not the one who memorizes one shiny button.

The limitation is confusion. Too many tools turn a film project into a software tasting menu. Pick a core stack. Add a supporting tool only when it solves a real problem. If you are signing up for platforms because you are anxious, take a walk. The movie does not need another login. It needs a finished scene.

Your Editing Suite: DaVinci Resolve and Final Cut Pro

All of this generation work is just raw material until you edit it, and editing is where a film actually becomes a film. DaVinci Resolve is my primary editing tool — it's free for the standard version, professional-grade, and includes the best color grading tools available at any price. I cannot overstate how important color grading is for AI-generated footage. Raw generations often have slightly mismatched color temperatures, inconsistent saturation, or that telltale "AI sheen" that signals synthetic imagery to a viewer's subconscious. A proper color grade unifies everything into a single visual world. Resolve handles this beautifully.

Final Cut Pro is my secondary editor, particularly for projects with tighter deadlines where Resolve's node-based color workflow would take too long. Final Cut's magnetic timeline is fast and intuitive. If you're on a Mac and you value speed over granular control, it's a solid choice. The secret weapon in both editors is the ability to layer. AI-generated footage, B-roll from Seedance, titles, sound design, music — all stacked and timed and blended into a coherent experience. Editing is not assembly. It is composition. It is rhythm and pacing and the felt sense of when to cut and when to hold. No AI tool can do this for you. This is entirely human work, and it matters more than any single generated clip.

Grumpy Old Man Tool Note: Final Cut Pro, Premiere, and Avid

Final Cut Pro, Premiere, and Avid all belong to the editing family, and each has its tribe. Final Cut is fast and friendly on a Mac. Premiere is everywhere and plays well with the Adobe crowd. Avid is the old professional battleship that still matters in traditional post houses. Pick the one that helps you finish, not the one that makes you feel important at parties.

For AI filmmaking, the editor is where you hide sins and create rhythm. You cut around bad hands, shorten weak motion, layer audio, insert reaction shots, build trailers, and make generated material feel intentional. Whether you use Final Cut, Premiere, Avid, or Resolve, the future skill is not button pushing. It is judgment. Where do you cut? Where do you hold? Where does the audience need to breathe?

The limitation is loyalty. People defend editing software like it is a family member. Stop it. The audience does not care what you edited on. They care whether the film works. Use the tool you can afford, understand, and finish with. The rest is barber-shop argument material.

Grumpy Old Man Tool Note: DaVinci Resolve

DaVinci Resolve is where the pile of AI material becomes an actual film. Before editing, you have clips. After editing, you might have a story. That distinction matters. Too many beginners think generation is the finish line. It is not. Generation is groceries. Resolve is where you cook the meal.

Resolve is useful for cutting, color grading, sound work, titles, pacing, and making footage from different tools feel like it belongs in the same universe. AI clips often come from different models with different contrast, grain, movement, and weird little moods. Resolve lets you unify them so the audience watches the film instead of wondering why every shot seems to have attended a different high school.

The limitation is learning curve. Resolve will not pat your head and tell you everything is easy. Good. Some tools should require respect. Learn the basics: cut, trim, audio levels, color, export. You do not need to become a Hollywood colorist by Thursday. You need to stop letting good footage die because you were afraid of the timeline.

Your Rough-Cut Assistant: Descript

Descript is an editor that works with text. It transcribes your dialogue, and you edit the video by editing the text — delete a word from the transcript and it disappears from the video. Rearrange sentences, the video rearranges with them. I use Descript for rough cuts and for removing filler words — "um," "uh," "you know" — automatically. For AI-generated dialogue, where every word is already precise, Descript is less essential. But if you're mixing AI voice with human recordings, it saves hours of timeline work.

Grumpy Old Man Tool Note: Descript

Descript is useful because it lets you work with audio and video through text, which is wonderful if your brain understands words before waveforms. You can cut interviews, narration, rough dialogue, and voice-driven material by editing the transcript. For a film with AI narration, interviews, explainer sections, or documentary-style structure, that can save a lot of time.

The possibility is rapid assembly. Generate voice in ElevenLabs, bring it into Descript, rough cut the structure, remove filler, rearrange lines, and then take the stronger version into Resolve or Final Cut for proper finishing. It is like sketching before painting. Nobody frames the sketch, but without it the painting often stinks.

The limitation is that Descript can make editing feel too much like word processing. Film still needs rhythm, pauses, looks, silence, and timing. Do not cut every breath just because the software lets you. Human beings breathe. Let them. Otherwise everybody sounds like they are being chased by a tax auditor.

Your Quality Enhancer: Topaz Video AI

AI-generated video sometimes needs a quality boost — upscaling from 720p to 4K, noise reduction, frame rate smoothing. Topaz Video AI handles this. I run final footage through it before festival delivery. The difference is subtle — cleaner edges, more stable motion, a slight increase in perceived production value. Think of Topaz as your digital film lab. You don't need it for every project, but for festival submissions and client work, that final pass is worth it.

Grumpy Old Man Tool Note: Topaz Video AI

Topaz Video AI is the cleanup crew. It helps with upscaling, sharpening, noise reduction, frame smoothing, and making footage look a little more finished. This matters because AI video can sometimes arrive looking close but not polished, like a suit that came back from the tailor one sleeve too long.

Use Topaz near the end of the pipeline, after you know the shot belongs in the film. Do not upscale every experiment like you are polishing rocks in the driveway. Pick the clips that will survive the final cut, then use Topaz to give them a cleaner presentation for festivals, clients, trailers, and online release. It is finishing work, not story work.

The limitation is overprocessing. Too much enhancement can make footage look waxy, crunchy, or fake in a new and exciting way. That is not progress. Use a light hand. The goal is to make the image feel more professional, not to make every face look like it was laminated at a copy shop.

Your Composers: Suno and Udio

AI music generation has reached a level where I no longer consider it a compromise. Suno and Udio are my two composers, and between them I can generate original scores in virtually any genre, tempo, and mood. The workflow is simple. I describe what I need — "somber piano with subtle strings, building to an emotional climax at ninety seconds, then resolving to quiet contemplation" — and the tool generates variations. I pick the one that fits. The quality is genuinely good. I've used AI-generated scores in festival-screened films and had musicians ask who composed them. Surprised, not offended. The music works. That's the only standard. For key emotional moments, I sometimes license a track from a human composer.

AI score for the bulk, licensed tracks for the peaks — the best cost-to-impact ratio.

Grumpy Old Man Tool Note: Suno and Udio

Suno and Udio are AI music tools, and they can give a filmmaker something that used to be expensive: original-sounding music direction without immediately hiring a composer. Music changes emotion faster than almost anything. Put the wrong music under a scene and you can turn grief into cheese in under eight seconds.

The possibility is temp scoring and mood exploration. Try a lonely piano theme, an uneasy synth bed, a corporate training jingle, a trailer pulse, or a strange folk song that sounds like it came from inside the movie's world. Use the AI score to discover what the film wants emotionally, then decide whether to keep it, revise it, license something, or bring in a human composer for the moments that really matter.

The limitation is rights, sameness, and taste. Do not assume every generated song is automatically safe for every use forever. Read terms. Keep records. And do not drown your film in music because you are afraid of silence. Silence is free and sometimes better. Amazing how often the cheapest tool is the one people forget.

How the Tools Connect

Here's the part that trips up most beginners. They look at this list — ChatGPT, Midjourney, HeyGen, ElevenLabs, Seedance, Runway, Sora, Kling, Pika, Luma, Veo, DaVinci Resolve, Descript, Topaz, Suno, Udio — and they feel overwhelmed. They think they need to master all of them before they can make anything. They don't. You need three tools to make your first film: something to write with, something to generate video with, and something to edit with. Start there. Add tools only when a specific project demands them. Do not subscribe to everything on day one. That way lies subscription creep, decision fatigue, and a drawer full of unfinished projects. My core workflow for a typical project looks like this.

ChatGPT or Claude for script development and dialogue refinement. Midjourney for character references and location packs. HeyGen for avatar performance. ElevenLabs for voice generation, connected to HeyGen through an API key so the cloned voices appear as options in the HeyGen interface. Seedance for B-roll, establishing shots, and atmospheric inserts. DaVinci Resolve for editing, color grading, and final delivery. Suno or Udio for score. Topaz for a final quality pass on the exported master. That's eight tools for a complete production pipeline. Some projects need fewer. Some need more — Kling for action sequences, Pika for effects, Luma for transitions, Runway for fixes. But those eight are my starting point for almost everything.

Grumpy Old Man Tool Note: Leonardo and the Budget Image Tools

Leonardo and similar lower-cost image tools matter because not everybody wants to start by paying for every shiny subscription on the internet. Good. Restraint is healthy. A free or cheaper image generator can help you test characters, locations, posters, and style before committing money to a bigger workflow.

The future use is early exploration. Build rough character portraits, quick environment references, social graphics, thumbnail ideas, or concept tests. Then take the best ideas into Midjourney, Seedance, Runway, or your editor. A cheap tool can be a scouting trip. You do not have to build the hotel just to see whether the town has a main street.

The limitation is consistency and polish. Budget tools may not give the same control, style coherence, or image quality as the stronger paid platforms. That is fine. Use them for what they are: idea generators. Do not confuse saving money with lowering standards. Save money where it does not hurt the film. Spend where it does.

The Combinations Nobody Is Talking About

Now let me tell you about the possibilities that keep me excited, because this is where your imagination matters more than any tutorial. HeyGen plus ElevenLabs plus Seedance equals a complete film studio for dialogue-driven work. But what about HeyGen plus Pika for a film where the characters exist inside a stylized, animated world? What about ElevenLabs plus Luma for a dream sequence where the narration guides the visual morphing? What about Kling's fight choreography cut together with Seedance's atmospheric B-roll for a thriller that has one action set piece surrounded by moody tension? Consider this: what if you used Pika's transformation effects not for a single scene but for an entire film's visual language?

A movie where reality slowly dissolves over the course of the runtime, the visual fidelity degrading and reconstituting in sync with the protagonist's psychological state. That is a film no traditional production could make economically. AI filmmaking makes it possible. Or this: what if you used Veo's synchronized audio generation for an entire scene shot from a single static camera position, letting the spatial audio do all the storytelling work? A character sits in a room, barely moving, while the sounds of the world outside tell us everything we need to know about what's happening. Radio drama with pictures. Pure cinema.

Or this — and this is the one that really excites me: what if you used Luma's keyframe interpolation to create transitions between memories in a documentary about someone's life? A photograph from childhood morphs into the adult standing in that same spot decades later. The visual continuity creates an emotional connection that no amount of narration could achieve. This technology opens doors that traditional filmmaking cannot even approach.

The Honest Truth About Limitations

I need to pump the brakes for a minute, because I promised you honesty and optimism in equal measure. These tools are remarkable, but they are not magic. Complex hand movements still challenge current systems — fingers have a way of looking wrong at exactly the wrong moment. Long continuous shots with multiple characters interacting physically are still difficult to generate coherently. And sustained emotional arcs that require dozens of subtle physical beats across a full performance are still better handled by real actors. The fix is not to give up on AI filmmaking. The fix is smart design. Lean into close-ups, where the face does the work. Lean into dialogue, where words carry the weight.

Lean into B-roll and atmosphere, where the tools are strongest. Use editing to cut around the limitations rather than trying to generate through them. Every filmmaker in history has worked within constraints. The difference now is that your constraints are different — not necessarily harder, just different.

What You Actually Need to Start

If you're feeling overwhelmed by the size of this toolkit, let me simplify. Here is the minimum viable stack to make your first film. ChatGPT Free for writing. One image generator — Midjourney if you can afford it, Leonardo free if you can't. One video generator — Runway or Seedance, whichever feels right to you. HeyGen for avatar performance. ElevenLabs for voice. DaVinci Resolve Free for editing. Suno or Udio free tier for music. That stack will cost you somewhere between zero and a hundred dollars a month, depending on which tiers you choose, and it is enough to produce work that can screen at festivals. The key is not having the best tools. The key is finishing something with the tools you have.

A completed film made with modest tools is infinitely more valuable than a perfect film that exists only in your imagination because you were waiting for the right subscription tier. In the next chapter, we're going to get practical. I'm going to walk you through setting up your workspace — not just the software, but the physical space, the habits, the mental framework that lets you produce consistently. Because tools are only as good as the person using them, and the person who wins is the one who shows up every day and does the work.

Chapter 3: The AI Film Lab That Fits on a Desk

I want to punch the next person who asks me what computer to buy.

Not hard. Just a little tap. Enough to get their attention. Because I've been asked this question forty times and the answer hasn't changed since the first time: the computer you already own is the computer you need. I've seen grown men spend three weeks comparing graphics cards online while their perfectly good laptop sits on the desk, gathering dust, wondering what it did wrong. Your machine is not the problem. Your willingness to start is the problem.

Chuck gets it. Chuck's a retired plumber I know from the hardware store, fixes everything in his house with a twelve-dollar wrench and whatever's in his basement. He told me once, "Jim, the best tool is the one you already own that you actually use." I think about Chuck every time someone shows me their color-coded spreadsheet of potential computer builds. The answer is: the one sitting on your desk right now, turned on, with a web browser that isn't more than two years stale. That's it. That's the magic machine.

Most of the heavy lifting in AI filmmaking doesn't happen on your computer at all. It happens in data centers owned by companies whose electric bills could fund a small nation. Let me say that again so it sinks in. Your laptop is a window. That's all. A window looking into a room full of very expensive servers doing math you don't understand. When you generate an image in Midjourney, create an avatar in HeyGen, or render a voice line in ElevenLabs, your machine isn't doing the work. It's watching someone else do the work. This means the threshold for entry is lower than almost anyone expects.

A five-year-old MacBook Air — the kind my wife uses to order gardening supplies — handles the entire web-based workflow without a hiccup. Writing, image generation, avatar creation, voice work, project management: all of it happens in your browser. Your computer is a window, not an engine.

The bottleneck shows up later, when you start editing video locally. DaVinci Resolve, which is the editing software I recommend and which happens to be free for most of what you'll need, does benefit from dedicated graphics hardware. For smooth 1080p editing, you want something with a discrete GPU and about 4 GB of video memory — an NVIDIA GTX 1650, a Mac with an M1 chip, or anything in that neighborhood. For 4K work or heavy color grading, you'd like 16 GB of system RAM and a GPU with 6 to 8 GB of VRAM. An RTX 3060 or Mac M2 Pro sits right in the sweet spot.

Anything beyond that is a luxury, not a requirement. The AI generation is rarely the problem. The editing is where hardware matters, and even then, you can work around limitations by generating at 1080p instead of 4K and using proxy media — smaller, temporary files that stand in for your full-resolution footage during the edit.

Before you spend a dollar on a new machine, do what I tell everyone to do. Open your Task Manager on Windows or Activity Monitor on Mac while you're working. Run your current workflow. Look at what actually maxes out. If it's your internet connection, a new computer won't help you one bit — call your ISP instead. If it's your RAM, a memory upgrade costs about fifty bucks and takes ten minutes to install. If it's your storage drive, an external SSD costs under a hundred dollars and will hold three or four short film projects.

Don't guess at your bottleneck. Measure it.

My neighbor replaced his entire computer because he thought it was too slow, then discovered the real problem was that his router was ten years old and couldn't handle the upload speeds. Bought a new router for eighty dollars, problem solved. The old computer was fine.

Here's what I tell people who call me in a panic about their setup: spend your money on tool subscriptions that produce footage. Upgrade hardware after you've proven to yourself that you'll actually finish projects. I've seen too many people drop two thousand dollars on a computer, use it for three weeks, then let it gather dust while they "research" the next upgrade. The best computer for AI filmmaking is the one that gets turned on every day.

What You're Actually Building

Let me tell you about Gloria before we get into folders and filenames, because Gloria is why any of this matters.

Gloria is seventy-six years old. She lives in a retirement community in Tucson with a pool she never uses and a patio that gets too much afternoon sun. Four months ago she started making a series called "Porch Stories" — five-minute episodes where her avatar interviews historical figures about their lives. Her avatar sits in a rocking chair on a digitally reconstructed veranda while Thomas Edison explains why he really stole the lightbulb, or Sacagawea describes the Missouri River in 1805. Gloria writes the scripts. Her avatar performs them. The backgrounds are generated from period photographs. Her "film lab" is a Chromebook on a TV tray.

She's produced twenty-three episodes. Her granddaughter posts them on TikTok. The last one got two million views.

This is what you're building. Not a folder structure. Not a subscription stack. A creative engine that produces work while you sit in your living room in sweatpants. The lab is just the container. Gloria is the point.

The Boring Part That Saves Your Life

Now let's talk about the part nobody wants to talk about: organization. Disorganized files kill more projects than bad prompts. When you can't find the right character reference, when your renders all have names like "final_v2_ACTUALFINAL_v3," when the perfect voice line you recorded last Tuesday is buried somewhere in your Downloads folder next to a PDF about refinancing your mortgage — you stop being a filmmaker and start being an archaeologist. Digging through layers of sediment trying to find the artifact you need.

Organization isn't bureaucracy. It's the infrastructure that lets you think about film instead of filing.

Create one master folder called PROJECTS. Simple, boring, obvious. Inside it, each film gets its own folder named with the project title and start date: FATHER_FREQUENCY_2026_01. The date isn't decorative. It prevents the chaos that happens six months later when you start the sequel, or the remake, or the version you have to start over because the first attempt taught you what not to do.

Inside each project folder, build your subfolders. Start with SCRIPTS — every draft from your first terrible vomit-pass to the locked shooting script. Add CHARACTERS for character bibles, reference portraits, wardrobe notes, and avatar training files. Each character gets their own subfolder. Add LOCATIONS for your setting reference packs, VOICES for voice lines organized by character and scene, STORYBOARDS for panels and shot lists, RENDERS for raw video clips organized by scene, EDIT for your DaVinci Resolve project file and cache, SOUND for dialogue and music and effects, and EXPORTS for review versions, festival deliverables, trailers, and the final master.

That sounds like a lot of folders. It is. This structure seems excessive until you're seventy files deep and need one specific voice line in under ten seconds.

Then it saves your sanity.

File naming is where creative people tend to lose their minds. A filename should tell you everything you need to know without opening the file: project abbreviation, content type, scene or subject, version, and date. Use underscores, not spaces — spaces break command-line tools and some web uploaders. A good name looks like: S01_SH01_CloseUp_v01_2026-06-17.mp4. That tells you Scene 1, Shot 1, a close-up, version 1, recorded on June 17th. A bad name looks like: marcus face final REAL FINAL.jpg — which tells you someone panicked and saved the same file three times with increasingly desperate adjectives.

Pick a naming convention on day one and enforce it like it matters, because it does. The five minutes you spend renaming a file correctly will save the thirty minutes you'd spend figuring out which "final" version actually made it into the edit.

The Arcade Economy

Now let's talk about money, because this is where the diner wisdom really kicks in. AI filmmaking runs on multiple currencies, and understanding them upfront saves you from the rude surprise of running out mid-scene. Think of it like an old arcade, the kind I used to take my kids to in the eighties. You walk in with a handful of cash, you get tokens at the change machine. Pinball costs one token. The racing game costs three. The boxing game costs five. Different games, different prices, same principle.

HeyGen credits are your arcade tokens for video. When you subscribe, you get a bundle of credits each month. Every video you generate costs some of those credits. A standard talking-head video runs about 20 credits per minute.

Avatar Shots — the cinematic, poseable version you'll use for dramatic scenes — burn credits much faster: 4 credits per second at 720p, 10 credits per second at 1080p. That means a 15-second Avatar Shot at high quality costs 150 credits, which on some plans is most of your monthly allowance. The discipline is to draft at 720p, get your pose and expression right, then render the keeper at 1080p.

Don't pay for ten passes when one will do.

The activities that cost zero credits are worth memorizing: creating a Digital Twin, cloning a voice, editing a project, writing scripts, previewing voices — you only spend when you hit Generate. Until that button gets pressed, experiment all you want for free.

ElevenLabs doesn't use credits. It uses characters. Yes, actual characters — letters, spaces, punctuation marks. The word "hello" is five characters. A normal sentence is fifty to a hundred. A minute of narration is roughly a thousand characters. Their free tier gives you about ten thousand characters, which translates to roughly seven minutes of speech. The Starter plan bumps that to thirty thousand, and the Creator plan gives you a hundred thousand — about seventy minutes of speech, which is enough to record every line of a short ensemble piece.

For a feature-length dialogue film, you'll want the Creator tier at minimum, because running out of characters halfway through your climactic scene is the kind of frustration that makes grown men consider taking up golf instead.

Here's the wrinkle that trips people up: when you connect ElevenLabs to HeyGen, you're running two wallets simultaneously. The video rendering pulls from your HeyGen credit budget, but the voice generation pulls from your ElevenLabs character budget. You can have plenty of HeyGen credits left while your voice goes silent because you've burned through your ElevenLabs characters.

Check both dashboards before a big project. It's like checking both your gas tank and your oil before a road trip. Basic maintenance.

The Guy at the Laundromat

While we're on the subject of subscriptions, let me tell you about the lead character.

the lead character works the overnight shift at a twenty-four-hour laundromat in Cleveland. Three in the morning, fluorescent lights humming, three washing machines running in sync. He has a phone, a pair of cheap earbuds, and a folding chair. That's his whole lab. He films a noir series called "Spin Cycle" — ten episodes so far — about a private detective who solves crimes at a laundromat. The hero is the lead character's avatar, fedora and all. The locations are AI-generated 1940s interiors. The rain on the windows is generated. The saxophone music is stock.

He writes during the slow hours between midnight and four. He generates footage on his phone during breaks. He edits sitting in that folding chair. His total equipment budget could be covered by the quarters in the change machine.

the lead character doesn't have a better computer than you. He has a better excuse to start.

Managing Your Stack

Subscription management is where I see the most waste. People sign up for everything on day one, use twenty percent of the features, and resent a hundred percent of the charges.

Start with the free tier. It's genuinely enough to produce a complete three-to-ten-minute short, provided you plan carefully and don't waste generations on experiments that don't serve the final cut. The constraints are actually good for you — when you only have fifty free image generations for your entire project, you think harder about each prompt. That discipline makes you a better filmmaker than unlimited generations ever will.

When you upgrade, do it because a current project is blocked without the paid feature, not because a YouTuber recommended it. The question isn't "would this be useful?" The question is "can I finish this week's work without it?"

Drop tools you haven't used in thirty days. Cancel without guilt. You can always resubscribe. I review my stack on the first of every month — every subscription, its cost, what it produced. Takes five minutes. Saves hundreds annually.

Before you make your first video in HeyGen, do one thing that will save you hours of frustration later: set up your Brand Kit. You'll find it in the Settings menu. This is where you teach HeyGen your colors, your logo, your fonts, and — this is the part most people skip — your pronunciation glossary.

The glossary is the most underused killer feature in the entire platform. Any proper noun, product name, technical term, or person whose name isn't pronounced the obvious way gets added there. If your film features a character named Siobhan, and you don't add the pronunciation guide, your avatar will say it like it's spelled, and every Irish person watching will wince. If your story involves a fictional pharmaceutical called Zynexor, the glossary is what keeps your narrator from turning it into gibberish.

This step takes five minutes and saves hours of re-recording later.

While you're in Settings, lock your default aspect ratio to whatever your project demands — 16:9 for YouTube and web, 9:16 if you're doing vertical content — and set your default voice once you have a clone you're happy with. Saves one click on every new scene, and those clicks add up.

The Recording

Now, the recording. This is the part that determines the quality of everything that follows, so pay attention like your film depends on it — because it does.

Your phone is the best camera for this job. Better than your laptop webcam, better than most dedicated webcams you could buy. Use the rear camera, not the selfie camera — it's higher resolution and doesn't apply beauty filters that smooth your skin into something the computer can't recognize as real.

Hold the phone in landscape orientation, sideways, propped on a stack of books or a small tripod.

Do not hold it in your hand. Nobody wants a seasick avatar.

Frame yourself from the chest up, with the camera at eye level, about two to three feet away. Too close distorts your face like a funhouse mirror. Too far makes your features too small for the model to read properly.

Lighting is the difference between an avatar that looks like a person and an avatar that looks like evidence from a crime scene. Face a window with soft natural daylight if you can — stand about four to six feet away so the light is even across your face. If you're using artificial light, put a lamp angled at your face so it fills the shadows under your eyes and chin. Avoid strong overhead light only — it creates dark shadows that make you look like you've been up for three days. Never put a window or bright light directly behind you — it turns your face into a silhouette, and the camera can't see you.

Think of it as taking a passport photo, but in motion. Neutral light, plain background, no sunglasses, hair off your face, looking at the camera. Now imagine that photo, but you're talking naturally for fifteen seconds. That's exactly what the computer needs.

The fifteen-second recording teaches HeyGen how you move, gesture, and hold yourself when you speak. This is critical: a flat, deadpan delivery produces a stiff, robotic avatar. The model copies the energy you put in. During your fifteen seconds, use your hands naturally, smile and frown, vary your tone, move your head a little, look directly into the camera. If you want your avatar to gesture naturally in its output, show hand gestures in the reference. If you want a calm, minimal-motion avatar, record a calmer reference.

The model picks up everything — your energy level, your posture, your rhythm.

Record three takes if you have time. The third one is almost always the best, once you've shaken off the awkwardness of watching yourself on camera.

Within those fifteen seconds, you need to speak your consent code. HeyGen generates this code uniquely for you when you start the avatar setup — it'll look something like "8472" or "HG-391." Speak it clearly within the first five seconds, at normal volume, using the recommended phrasing: "For safety purposes, my unique code is [code]."

This isn't some corporate hoop to jump through. It's the difference between a legitimate Digital Twin and a deepfake. Without this protection, anyone could upload a fifteen-second clip of any person — a celebrity, a politician, your neighbor — and create a synthetic version of them without permission. The consent code exists because the person being cloned must be physically present at the recording and speak the code themselves. You can't reuse old codes. You can't fake it with text-to-speech. The system rejects uploads where the code isn't detected clearly, and there's no penalty for re-recording — it simply refuses to proceed until it hears the code properly.

This is what makes the whole system ethical. Don't resent it. Respect it.

Your voice sample is equally important, and it's recorded separately in ElevenLabs. Find a quiet room — a bedroom or office with the door closed. Not the bathroom, even though the acoustics might sound okay to you. Bathrooms have hard surfaces that create reflections and weird resonances the voice model will learn and reproduce.

Hold your phone six to eight inches from your mouth. Too close and you get popping sounds on P and B words. Too far and the room noise becomes part of your clone's permanent personality.

On iPhone, open Voice Memos, tap the settings gear, and set Audio Quality to Lossless — not Compressed, which throws away the subtle breath and tonal details that make a voice sound human. On Android, set your recording quality to the highest bitrate available, typically 256 kbps or higher.

Talk for one to three minutes about anything — what you did yesterday, a vacation memory, what your business does. The content doesn't matter. What matters is that you sound natural, conversational, alive. Vary your tone. Pause naturally. Let some words have more energy than others. A flat, stiff recording produces a flat, stiff clone. The computer learns your delivery style, not just your pitch and timbre.

If you sound like you're reading a tax form, your clone will sound like it's reading a tax form forever.

Three Labs, One Blueprint

Here's where I want you to stop and think about what you're actually building. Most people read a chapter like this and picture one project sitting in one folder. One short film. That's fine. That's where I started. But the lab you're building right now — this folder structure, these subscriptions, these naming conventions — this infrastructure supports an entire creative practice, not just one film. And I mean creative practice in the wildest sense.

Think about Diego. Diego lives in a 1972 Airstream trailer and makes documentary films about forgotten American towns. His entire lab fits in a rolling suitcase: a laptop, a portable light, a phone on a mini tripod. He interviews historians and locals in the morning, generates period-accurate backgrounds of what those towns looked like fifty years ago in the afternoon, and edits episodes at campgrounds with spotty WiFi. His avatar — trained from recordings he made in a motel bathroom with a towel stuffed under the door — narrates over archival photos and AI-generated reconstructions.

He's produced seventeen episodes. From a trailer. With a view of the Rockies out his window.

Or picture Priya, who makes one sixty-second film every single day. One. Every day. Her lab is deliberately microscopic: three avatar characters, two locations, one voice. She writes her script in the morning while her coffee brews, generates her shots during her lunch break, and edits while her kid does homework in the next room. Her file structure is so clean you could eat off it. Her naming convention fits in a tweet. She has 340 films now. Three hundred and forty. Most of them are small — a conversation, a revelation, a tiny human moment. But the cumulative weight of that practice is staggering. She's become a better filmmaker in one year than most people become in ten, because she ships daily.

Now imagine four comedy writers in a shared workspace in Chicago who pooled their resources into one collaborative lab. They share a stable of twenty avatar characters — politicians, absurdists, straight men, eccentrics — trained from their own faces and voices, cross-cast depending on who wrote which sketch. Their PROJECTS folder has a shared CHARACTERS directory where anyone can pull a pre-trained avatar for a new script. They have collaborative naming conventions, shared pronunciation glossaries, a group calendar for subscription renewal dates. On Monday mornings they read scripts aloud and argue about which avatar should play which role.

They produce four ten-minute episodes a week, rotating who directs, and they've built an audience of fifty thousand people who have no idea the "ensemble cast" is actually four guys in a room arguing about whose turn it is to buy coffee.

The same lab. The same infrastructure. Wildly different outputs.

That's what you're building here — not a tool for one film, but a studio that can produce whatever your imagination comes up with. A mobile documentary unit. A micro-content factory. A collaborative writers' room. Or something nobody's thought of yet. The lab doesn't judge. It just works.

Backing Up Your Life's Work

The final piece is your backup system, and I'm only going to say this once: if a file exists in only one place, it does not exist.

Your working drive will fail. It's not a question of if. It's a question of when. Maintain at least two copies — one local, one in the cloud. A second external drive synced weekly covers your local backup. A service like Backblaze — about seven dollars a month for unlimited backup — handles the offsite copy. For your project files specifically — the scripts, the Resolve project, your asset registry — use cloud sync as primary storage. These files are small and precious. For media files — video renders, voice lines, music — the size makes cloud sync impractical, which is why a background backup service is worth every penny.

Compare seven dollars a month to the cost of re-rendering your entire film because the files disappeared. It's the cheapest insurance you'll ever buy.

One last thing about cloud sync: it is not backup. Google Drive, Dropbox — these synchronize your local files to the cloud. If you accidentally delete a file locally, that deletion syncs upward. If something nasty encrypts your local drive, the encrypted versions sync upward too. True backup keeps previous versions independently of your current files. Use a dedicated backup service, or manually maintain a separate copy that doesn't automatically mirror your deletions.

Grumpy Old Man Tool Note: Google Drive and Dropbox

Google Drive and Dropbox are not glamorous, which is why they are useful. Nobody brags at a party about folder structure, but messy files have killed more small projects than bad reviews.

These tools keep scripts, renders, references, contracts, exports, and client files in places where you can actually find them instead of digging through a desktop that looks like a raccoon nest.

Use cloud storage as the spine of the project. Script drafts in one folder. Visual references in another. Generated clips labeled by scene. Audio takes named properly. Final exports separated from experiments. Pair this with Google Docs or Notion for the show bible and you start building a production system instead of a junk drawer with Wi-Fi.

The limitation is that sync is not the same as backup. If you delete garbage locally and the cloud helpfully deletes it everywhere, congratulations, technology just made your mistake more efficient. Keep separate backups. Use external drives. Version important files. A filmmaker without backups is just a future sob story.

The discipline is mechanical and boring, which is exactly why it works.

Set up the folders once. Name files correctly every time. Back up at the end of every session. These habits don't feel creative, but they create the conditions under which creativity survives production chaos. The filmmakers who finish aren't always the most talented.

They're the ones who can find their files.

Don't build the perfect folder structure before you have a project. Create the PROJECTS folder and one project subfolder. Build the rest as you generate assets. An empty, elaborate folder structure is procrastination dressed up as preparation. A populated, imperfect structure is a film in progress. The lab exists to serve your creative work, not the other way around.

Now that it's built, it's time to think about how to work inside it. That's the next chapter — and it's the one that matters most.

Chapter 4: The Director and the Technician

My neighbor across the street spent two years building his own deck. Every weekend, hammer in hand, cursing at crooked boards, buying tools he'd use once and never touch again. The deck is fine. It's a perfectly adequate deck. But my other neighbor — the smart one — hired a contractor, picked the stain color he wanted, and had the whole thing done in three weeks while he was at work. Both men have decks. One of them also has a finished basement, a repaved driveway, and a vacation album from the ten weekends he didn't spend arguing with lumber. That's the difference between a technician and a director. And if you're making AI films, you need to decide right now which one you are.

When I started NJ Film Productions, I thought my job was to learn every tool and execute every technical step myself. That's a technician's mindset, and it will kill your productivity faster than bad lighting or a subscription you forgot to cancel. A technician asks, "How do I do this?" A director asks, "What do I want, and how do I get something else to do it for me?" The difference is not subtle. It's the difference between the man who builds his own house with his own two hands and the man who hires architects, contractors, and inspectors to build it to his specifications. Both can produce excellent houses.

But the second man can build ten houses in the time the first builds one, and each of those ten houses benefits from expertise the generalist will never possess. Your AI tools are not your hands. They are your crew. When you write a prompt, you are not typing magic words into a machine. You are giving instructions to a digital cinematographer, a digital gaffer, a digital production designer, and a digital visual effects artist who happen to live inside the same interface. The quality of your output depends on the clarity of your direction, not your ability to manually operate every subsystem. Think about a traditional film director on a physical set.

That director may not know how to adjust the intensity on a specific light or how to calibrate a camera's color science. But the director knows exactly what the scene should feel like, and can communicate that feeling in terms the specialists understand. "I want her face to feel isolated, like she's the only person in a room full of people." The gaffer hears that and creates a single pool of light surrounded by darkness. The cinematographer pushes in slowly on a longer lens. The sound designer hollows out the ambient noise so her voice floats in dead space. Your prompts are that direction. Your taste is that clarity. The tool does not replace your judgment — it executes your judgment at speed.

The most expensive mistake in AI filmmaking is entering a prompt before you know what you want. When you type "cinematic shot of a woman in a room" into a video generator, you have already failed. You have outsourced the creative decision to an algorithm that does not know your story, your character, or your emotional intent. You will get something that looks impressive but means nothing. A director knows what they want before they ask for it. This does not mean visualizing every pixel in advance. It means you can answer four questions for every shot: What does the viewer need to know after seeing this? What should the viewer feel? Where is the visual weight — what draws the eye first?

How does this shot relate to the one before it and the one after it? If you cannot answer those four questions, you are not ready to prompt. Go back to your storyboard, your script, or your character notes until the answers are clear. The tool will wait. Your creative energy will not replenish if you spend it generating random variations hoping one accidentally works. For the sample short, I knew before I generated a single frame that the protagonist's apartment needed to feel like a cocoon of grief slowly becoming a prison. The warm lamp light that felt comforting in scene one needed to feel suffocating by scene six. I knew the emotional map. The prompts were just translations of decisions I had already made.

Now here's the problem that keeps me up at night, or would if my back let me sleep through the night in the first place: AI filmmaking has almost no natural constraints, and that freedom is paralyzing. Traditional filmmaking forces decisions on you. You can only shoot when the location is available. You can only afford so many lighting setups. Your actor can only do so many takes before exhaustion sets in. These constraints are frustrating, but they are also protective. They force you to decide and move on. AI filmmaking has none of that. You can generate a hundred variations of the same shot. You can change the lighting, the angle, the color palette, the time of day, and the weather conditions with a few keystrokes.

You can try ten different styles before lunch. Every additional option consumes a unit of your limited daily decision-making capacity, and by the end of the afternoon, your judgment is so depleted that you accept mediocre output simply because you cannot evaluate it anymore. Psychologists call this decision fatigue, and it is real. A traditional film director might make fifty mean- ingful decisions in a day. You, operating a one-person AI studio, can face fifty meaningful decisions in a single hour, and then fifty more, and then fifty more, until you are circling the same shot for the third hour wondering if version sixty-three is marginally better than version sixty-two.

I have watched cre- ators spend entire afternoons generating the "perfect" establishing shot. The shot they finally chose was rarely better than the third or fourth option they generated. The difference was barely visible. But their decision-making capacity for the rest of the day was destroyed. The solution is not willpower. Willpower fails when you are tired. The solution is structure — systems that remove trivial decisions before they reach you. The most productive AI filmmakers I know work with deliberate constraints. They decide the rules before they start, and they follow those rules until the project is done.

Lock your color palette before you generate — choose three to five recurring colors, write them down, and every prompt references those colors. You no longer decide color on a shot-by-shot basis. That decision was already made. Pick one aspect ratio for the whole project and do not change it because one shot "might look cool" in a different format. Set a generation budget — I use a three-attempt rule. Generate three versions, pick the best one, move on. If none work, the problem is not the prompt. The problem is that you are asking the tool to do something it cannot do yet, and you need to redesign the shot.

Find one or two reference films or photographers that capture the look you want, and when you are unsure about a shot, ask yourself: "Does this feel like it belongs in the same world as my references?" If the answer is no, adjust or discard. Batch similar decisions — do all your storyboards before you generate any video, lock all your character voices before you record dialogue, choose your music direction before you start the final mix. Context-switching between creative tasks burns decision energy faster than working in batches. These constraints feel limiting for about ten minutes. Then they become liberating. You stop second-guessing yourself because the parameters are already set.

Your mind is free to focus on the creative questions that actually differentiate your film from everyone else's. Your first project should fit in a shoebox. One room. One character. One minute. I am not being cute. I am being practical. The point of your first project is not to make something impressive. The point is to prove your pipeline works — to go from script to storyboard to generation to edit to sound in one continuous workflow, and to finish. A one-minute film with one character in one room teaches you more about AI filmmaking than a twenty-minute epic that collapses under its own weight in week three. You learn how your avatar performs. You learn how your voice clone sounds. You learn how the edit rhythm feels.

You learn where the friction lives in your workflow. That knowledge scales to every project after. My first AI-assisted scene was exactly this: one weekend, one character, two minutes of dialogue. The lighting was inconsistent. The voice was slightly robotic. The editing was rough. But I learned my pipeline, and that knowledge was worth more than a perfect scene would have been, because it scaled to the sample short and everything after it. The iteration mindset is what separates people who finish films from people who accumulate folders of disconnected pretty pictures. Every generation is a rehearsal, not a performance. When a take doesn't work — and most won't — you are not failing. You are rehearsing.

Theater actors do twenty rehearsals for every performance. Film crews do multiple takes for every shot in the final cut. Why should AI filmmaking be any different? The difference is that your "rehearsals" happen in seconds instead of hours, and they cost fractions of a dollar instead of thousands. The iteration mindset means treating each generation as information. This prompt produced this result. The result tells me something about what the tool can and cannot do at this moment. I adjust. I try again. Each loop teaches me. The filmmaker who learns from every generation improves faster than the filmmaker who generates randomly hoping for a happy accident.

This brings us to the hardest lesson for creative people, and the one my wife had to beat into my head over thirty years of marriage: good enough for the job is not a compromise. It is a discipline. Eighty-five percent excellent and published today beats ninety-five percent and published next month, every single time. A finished imperfect film is more valuable than a perfect unfinished film in every possible dimension. It teaches you more. It builds your portfolio. It can be released, screened, criticized, and learned from. An unfinished project teaches you only what it feels like to abandon something. The perfectionism trap is especially seductive in AI filmmaking because the tools make endless iteration so easy.

There is always one more prompt variation to try. There is always a slightly better generation just around the corner. That promise is a lie. Perfection in creative work does not exist. There is only the point at which additional effort produces diminishing returns, and the point at which the work is complete enough to serve its purpose. I measure progress by a simple standard: is the work getting closer to finished, or is it getting closer to perfect? These are not the same direction. Getting closer to finished means making decisions, accepting trade-offs, assembling the pieces, and moving to the next stage.

Getting closer to perfect means circling the same decisions indefinitely, generating alternatives that are two percent different, and convincing yourself that you need just one more day. The brutal truth is that most of what you are perfecting will not be noticed by your audience. They will notice whether the story holds them. They will notice whether the characters feel real. They will notice whether the ending lands. They will not notice the background detail you spent three hours adjusting. At NJ Film Productions, we ship. That does not mean we ship garbage. It means we set quality standards at the start of a project, we work to meet those standards, and when they are met, we finish.

If the lighting in shot twelve is ninety percent of what I envisioned instead of one hundred percent, I move on. The film is not about shot twelve. It is about the cumulative experience of all the shots, the sound, the music, the performances, and the story working together. One slightly imperfect shot in a strong film is invisible. A perfect shot in a film that never gets finished does not exist. Not everything you make needs to be a masterpiece. In fact, insisting that everything be a masterpiece is the fastest route to making nothing. You need two categories of work: practice projects and portfolio pieces. A practice project exists to teach you something specific. Maybe you are learning a new video generation tool.

Maybe you are testing whether a particular visual style works. The goal is not a finished film worthy of public release. The goal is learning, and learning has its own quality standard: did I understand this tool or technique well enough to use it intentionally next time? Practice projects should be small, fast, and disposable. One scene. One character. One location. One minute. Do them quickly, evaluate them honestly, and then either apply what you learned to a portfolio piece or do another practice project targeting the next skill gap. A portfolio piece is work you would show to a potential client, a festival programmer, or an audience.

It has higher standards and a longer timeline, but those standards are defined in advance, not allowed to expand indefinitely. The danger is conflating the two. Treating a practice project like a portfolio piece means spending weeks on something that was never meant for release. Treating a portfolio piece like a practice project means shipping work that damages your reputation. Decide which category a project belongs to before you start, write that decision down, and hold yourself to the appropriate standard. When it is time to review your work — and you must review it, regularly and honestly — I use four passes, each one looking for something different. The Structure pass asks: does the story work from beginning to end?

Can a viewer who knows nothing follow the narrative, understand the characters' motivations, and experience an emotional arc? The Dialogue pass asks: does every line sound like something a human would say? Is there subtext, or is everything on the nose? The Renderability pass asks: which shots are giving me trouble, and is the problem the prompt or an impossible ask? Some things current tools simply cannot do well, and knowing the difference between a bad prompt and a bad tool saves hours. The Cut or Combine pass asks: which scenes are unnecessary, which can be merged, and where is the pacing dragging? These four passes, done in order, prevent the review from becoming a free-for-all where you notice everything at once and fix nothing.

Structure first. Always structure first. A beautiful render of a scene that doesn't belong in the film is still a scene that doesn't belong in the film. The four passes become a repeatable workflow, and a repeatable workflow is what turns occasional dabbling into a serious practice. Every project follows the same path: concept, script, storyboard, generation, assembly, sound, color, export. The tools within each step may change. The sequence does not. When you have a workflow, you know where you are at any moment. You know what comes next. You know when you are stuck and what moving forward looks like.

Without a workflow, every project feels like starting from zero, and the mental overhead of figuring out what to do next drains the energy you need for creative decisions. Write your workflow down. Refine it after every project. The workflow you have is better than the perfect workflow you keep planning to create someday. Now I want you to stop and really think about what you are becoming. Because what I'm describing — one person acting as writer, director, producer, cinematographer, editor, sound designer, and distributor — this is unprecedented in the history of film. Orson Welles couldn't do this. Spielberg can't do this. Every great filmmaker in history worked with crews, with studios, with budgets that required compromise and collaboration.

You don't need any of that. The constraint of working alone isn't a limitation anymore. It's a creative catalyst. It forces you to find solutions that crews never discover because they don't have to. It pushes you toward intimacy, toward precision, toward the kind of personal vision that gets diluted when twenty people all have opinions. Think about the genres that are perfect for solo production and barely anyone is exploring. Intimate character studies — two people in a room talking, the kind of film that lives or dies on the writing and the performances. Philosophical dialogues — a single speaker working through an idea, the kind of content that builds obsessive audiences who feel like they're in conversation with the creator.

Personal essay films — your story, your voice, your perspective on the world, delivered directly to the viewer without the filters of traditional documentary production. Experimental hybrid documentaries — mixing archival footage, AI-generated reconstructions, and avatar narration to tell true stories in ways that have never been attempted. Serialized micro-dramas — five-minute episodes released weekly, building an audience over time like a novelist releasing chapters. These are not compromises forced on you by limited resources. These are forms that didn't exist before because the economics of traditional filmmaking made them impossible. One person, one laptop, one vision, direct to audience. That's not a diminished version of filmmaking.

That's a new kind of filmmaking entirely. And here's what really excites me, the thing that keeps me at my desk even when my back is acting up: we are in the very first moment of this. The tools you are learning to use today will be better in six months, and unrecognizable in two years. But the mindset — the discipline of the one-person studio, the clarity of the director who knows what they want before they ask for it, the taste that develops through honest self-assessment — that mindset compounds. Every film you finish makes you a better filmmaker. Every review pass sharpens your eye. Every shipped project builds your confidence and your audience.

The filmmaker who starts today, with imperfect tools and a modest first project, will be ahead of the filmmaker who waits for the "perfect" toolset that never arrives. The constraint of starting now, with what exists, is what builds the judgment you'll need when the tools can do anything. You are not a button-pusher. You are a director. The button is just how you give instructions. What matters is what you want, why you want it, and whether you have the discipline to finish what you start. The lab is built. The tools are waiting. The only question is whether you'll direct them toward something worth watching, and whether you'll keep directing until the work is done. That's the mindset. Everything else is just technique.

Chapter 5: The Logline and the Leap

I have a neighbor, Frank, who can talk for twenty minutes about what he had for breakfast. Eggs, toast, the particular brand of jam his wife buys, how the toaster broke last Tuesday and the guy at the hardware store said it would cost more to fix than replace. By the time he gets to the coffee, I have mentally moved to another state. Frank is a good man. He cannot tell a story to save his life.

Most filmmakers, when you ask what their movie is about, sound like Frank. They ramble. They explain themes. They tell you about the symbolism of the rain, the influence of Tarkovsky, the personal trauma that inspired the third act. What they cannot do — what they have somehow never practiced doing — is sum it up in one sentence that makes a stranger lean in and say, "Tell me more."

That sentence is the logline. And for the AI filmmaker, it is not a marketing exercise. It is the foundation everything else stands on.

A logline is a single sentence, ideally under forty words, that names your protagonist, drops them into a situation, gives them a concrete goal, throws an obstacle in their path, and tells us what happens if they fail. Five elements. Strip one out and the whole thing goes wobbly. The protagonist needs to be specific — "a widowed sound engineer" beats "a man" every time. The goal needs to be something you can photograph — "to recover the stolen hard drive before his deadline" works; "to find peace" does not, because peace looks like nothing on screen. The obstacle needs to be active, not internal. The stakes need to be specific enough to hurt.

For the sample short, my logline was: "A grief-stricken sound engineer discovers his dead wife's unfinished recordings and must complete her final album while learning to let her go." That is twenty-two words. It tells you who, what, where, why, and what happens if he fails. Every scene I wrote after that, every prompt I fed into every AI tool, every decision about where to spend money and where to save it — all of it traced back to that sentence. The logline is your compass. Without it, you wander.

Here is what I do when I am developing a new idea. I take my raw notion — the thing I would ramble about at a party, Frank-style — and I paste it into an AI tool with a very specific prompt. I ask for eight different loglines, not three, not five, but eight. I want variety across all five elements: change the protagonist trait, shift the goal, swap the obstacle, raise the stakes. The first three will be competent and forgettable, like vanilla ice cream at a diner. Versions four and five start getting interesting. But version six, almost every time, version six has something I would not have thought of myself — an angle, an irony, a combination of elements that makes me sit up straighter in my chair. That sixth logline is where the gold lives.

When I was developing the sample short, the project was just an image: a man with headphones, staring at a waveform. The AI's first three loglines were functional — grieving father, mystery, search for truth. The sixth introduced a different angle entirely: what if the voice is not a gift but a trap? What if following these broadcasts leads him somewhere he cannot bear to go? That unexpected turn became the seed for the second act. The AI did not write my film. It proposed an idea I would not have considered, and that idea made everything better.

But here is the rule: never accept the AI's first suggestion. The first response from any model is statistically the most obvious one — the answer calculated to satisfy the broadest range of people asking similar questions. It is vanilla. Accept it and move on, and you will build an entire film on sand. By the time you realize the core idea is flat, you will have a synopsis, a beat sheet, scene outlines, and maybe even rendered footage that all point in the wrong direction. Fixing the foundation means rebuilding everything above it.

Instead, argue with the machine. Paste the logline you like best back into the tool and ask what is wrong with it. Ask it to play a cynical studio executive who has read ten thousand of these things. Ask it to tear your logline apart — what is predictable, what is vague, what feels like a movie we have already seen. The AI, freed from being helpful and polite, will flag weaknesses you missed. "The goal is not concrete enough to photograph." "The obstacle is internal, so the second act will be static." These notes save weeks of work. Then ask for three alternative angles that would surprise even that cynical executive. One of them might become your film.

Once the logline is locked — and I mean locked, not "pretty good, I will probably change it later" — you are ready for structure. And I want to be very clear about what I mean by structure, because the word makes some people nervous. They think structure means formula. They think it means cookie-cutter. They are wrong. Structure is engineering. It is the load-bearing walls that keep your story from collapsing in the middle.

The three-act structure is the simplest durable framework ever devised. It predates cinema by thousands of years. It is in Greek drama, Shakespeare, comic books, and every commercial film you have ever loved. A story has a beginning, a middle, and an end. The beginning sets things up. The middle makes things worse. The end sorts them out. Act One is roughly the first 25 percent of your runtime. Act Two is the middle 50 percent. Act Three is the final 25 percent. These are not rigid — a thriller might compress Act One and stretch Act Two, a character drama might linger in the setup. But the proportions exist for a reason.

They reflect how long an audience will wait before the story needs to escalate, and how much escalation feels satisfying before resolution.

For the AI filmmaker, though, structure is not just storytelling. It is production planning. Every beat you write translates into visual requirements: locations, characters, props, lighting conditions, weather, camera movements, special effects. In traditional filmmaking, you can write "a crowd of ten thousand watches as the spaceship crashes into the stadium during a thunderstorm" and figure out later how to shoot it. In AI filmmaking, that sentence is a suicide note. The AI cannot creatively frame its way around what it cannot generate consistently. When you structure your story, you are making decisions about render cost at the macro level.

This is where the fifteen-beat method comes in. I use it because it gives me the most practical plotting tool I have found. Blake Snyder popularized it through his Save the Cat books, and some writers find it constraining. I find it liberating. The beats are not a prison. They are a scaffold. You can build whatever you want on top of them, but without the scaffold, your story sags in the middle and possibly collapses at the end.

The fifteen beats are: Opening Image, Theme Stated, Setup, Catalyst, Debate, Break Into Act Two, B Story, Fun and Games, Midpoint, Bad Guys Close In, All Is Lost, Dark Night of the Soul, Break Into Act Three, Finale, and Final Image. Each one has a job. The Opening Image establishes tone and gives us a visual metaphor for the story's theme. In the sample short, it is the lead character alone at 3 AM, surrounded by mixing boards and shelves of vinyl records in his home studio, a framed photo of his wife turned slightly away — so he can see it, but not quite face it. That image tells us everything before anyone speaks. The Theme Stated comes early, usually from a supporting character who does not realize they are naming the film's central idea.

the lead character's sister tells him, "You can't keep her alive through those recordings, the lead character. Eventually you have to let the song end." The title pays that off.

The Setup establishes the world, the relationships, the flaws, what is missing. The Catalyst is the inciting incident — the event that breaks the protagonist's routine and poses the question that launches the main action. The Debate is hesitation, the last attempt to stay safe before committing to the journey. Break Into Act Two is the point of no return. The B Story carries thematic weight — in the sample short, it is the lead character's relationship with his sister a supporting character, the character who represents the theme and needs her brother present more than she needs her sister-in-law's ghost. Fun and Games delivers on the premise's promise — the trailer moments, the mystery, the treasure hunt. The Midpoint is a false victory or false defeat that raises the stakes.

Bad Guys Close In is where opposition mounts, internal and external. All Is Lost is the lowest point. Dark Night of the Soul is the emotional aftermath — the sitting with grief before the final push. Break Into Act Three is the new insight, the resolve to finish. The Finale resolves both the A story and the B story simultaneously. The Final Image mirrors the Opening Image, showing us the change.

Now here is what makes this method indispensable for AI filmmaking: the AI Complexity column. I add a column to my beat sheet that rates each beat by production difficulty on a scale of one to five. A one means interior, two characters, static camera, controlled lighting — something the AI can generate reliably in an afternoon. A five means crowds, complex action, animals, weather effects, extended continuous shots — the kind of thing that will eat your life and maybe your sanity.

the sample short keeps almost every beat at complexity one or two. The Fun and Games sequence scores a three because it requires multiple locations, but they are simple interiors and exteriors — no crowds, no weather, no action. The cemetery finale scores a three because it is an exterior, but it is a controlled exterior with two characters and stable lighting. This is a producible film. If more than 20 percent of your beats score four or five, your story is too expensive for a first project. Restructure it now, not after you have spent weeks trying to render the impossible.

The beat sheet with that AI Complexity column becomes your render budget planner. Identify your complexity clusters — three consecutive beats that all score four or five, and you have created a production bottleneck. You will spend disproportionate time on that section, which creates two risks: you might burn out and abandon the project, or you might rush through and accept lower quality just to get past the difficult stretch. Either outcome is bad. Restructure to separate your high-complexity beats with simpler material that gives you recovery time. Think of it like spacing out the hills on a road trip. You need flat stretches between the climbs.

Use the location column to batch your work. Every location change requires new environment prompts, new style references, and new consistency checks. If your beat sheet shows you bouncing between twelve locations throughout the film, you are working against yourself. Group scenes by location where structurally possible. Shoot all the apartment scenes together, even if they appear in different acts, because you can reuse the same environment setup. Work in location order, not story order. Keep a detailed log of which prompts and reference images produced the best results for each location so you can replicate them when you return to that environment later. This discipline saves hours.

The AI Complexity column does more than warn you about trouble. It helps you invent new formats. Once you know exactly what each structural element costs to produce, you can start playing with form in ways that traditional filmmakers cannot. Could you write a two-minute AI film that goes viral on social media? Absolutely — compress the three-act structure into 120 seconds, make your setup fifteen seconds, your confrontation ninety, your resolution fifteen. Could you make a ten-episode micro-series where each episode is sixty seconds? Yes — treat each episode as one beat from a larger structure, so the ten episodes together form a complete story arc. Could you create an interactive story where viewers choose the path?

The beat sheet becomes a decision tree — at each beat, the viewer's choice determines which branch activates next, and you render only the beats on that branch. An anthology film where each segment uses a different avatar style? The AI Complexity column tells you exactly which segments are cheap and which are expensive, so you can budget your creativity accordingly.

The three-act structure is not just for features. It scales down to thirty-second ads and up to ninety-minute films. It is architecture. And architecture does not care about size — it cares about load-bearing capacity.

When it comes to actually writing the screenplay, there is one rule every filmmaker learns eventually: one page equals approximately one minute of screen time. A five-minute short needs about five pages. A ninety-minute feature needs about ninety pages. This is not precise — dialogue-heavy scenes run longer on the page than action scenes, because a page of two people talking might take two minutes to perform while a page of "the car explodes" takes three seconds. But the rule keeps you honest about scope. If your short film script is thirty pages, you are not making a short film. You are making a feature that you have not admitted to yourself yet.

Writing for AI voice requires a different ear than writing for the page. Punctuation controls pacing. Periods create a short pause and a tone reset. Commas create a brief beat. Ellipses create a longer hesitation, a trailing off. Question marks raise the voice at the end. Exclamation marks add emphasis. These are not decorative flourishes — they are timing instructions. The voice engine reads them the way a musician reads notation. Write a run-on sentence with no commas and the voice will barrel through it like a train without brakes. Write fragments. Short sentences. Then a longer one that flows and builds and carries the thought to its natural resting place. Vary the rhythm. Read your script aloud.

If you sound bored reading it, the AI will sound bored speaking it.

Audio tags — [whispers], [angry], [sighs] — are seasoning, not the whole meal. One or two per sentence is usually enough. Stack five tags on a single line and you confuse the model. Use them at the point where you want the effect to start, and understand that the tag persists until a contrasting tag appears or until a natural sentence boundary. [Whispers] at the beginning of a line means everything after it is soft, intimate, close to the ear. If you want the character to return to normal volume, you need a new tag — [normal], [firm], whatever matches the emotional register — or a sentence break. Think of audio tags like salt. A little brings out the flavor. Too much ruins the dish.

When it is time to move from screenplay to HeyGen script, the conversion process strips away everything except what the avatar needs to speak. Scene headings come off — HeyGen handles visuals through the avatar, background, and Looks system. Action lines come off — the avatar cannot perform them. What remains is the spoken dialogue, assigned to each character, with audio tags added for emotional direction and pauses where timing needs control. Each character gets their own separate script block. In HeyGen, you generate one character's video, then the other, and edit them together.

For true two-character shared-frame scenes with real eye contact, you use Avatar Shots, which is cinematic and beautiful and more expensive — reserve it for the moments that matter most, the climax, the reveal, the emotional handoff. For everything else, cutting between speakers works fine and costs less.

Word counts matter here because timing matters. Roughly fifty words equals thirty seconds of screen time. One hundred words equals one minute. Two hundred words equals two minutes. Five hundred words equals five minutes. This is approximate — some voices speak faster, some slower, pauses add time — but it is close enough to plan with. When you are budgeting a scene, count the words. When you are planning a film, add up the word counts. The math keeps you honest.

Structure is not theory. It is engineering. The logline is your blueprint. The beat sheet is your construction schedule. The AI Complexity column is your budget estimate. Every decision you make at the story level determines what you will spend at the production level — in time, in money, in iterations, in frustration. Plan tight. Build strong. The tools will try to tempt you into skipping steps, into discovering the story as you go, into letting the AI figure it out. Do not listen. Structure is your job. Rendering is the AI's job. Know which is which, and you will finish films while other people are still rearranging their beautiful, useless shots in folders marked "draft 47."

In the next chapter, we are going to talk about the people who inhabit your structured story — how to build characters who want things, fear things, hide things, and change by the end. Because a perfectly structured film full of cardboard cutouts is still a failure. The structure holds up the building, but the characters are why we walk inside.

Chapter 6: Characters, Dialogue, and the AI Writer's Room

My wife has a saying about people who talk a good game but have nothing underneath: "All icing, no cake." She says it about politicians, about the guy who sold us a car with a transmission that lasted eleven days, about my brother-in-law's elaborate business plans that never seem to produce actual income. I have learned over forty-odd years of marriage that she is usually right, and I have learned over twenty-eight years of filmmaking that the same rule applies to characters.

A character is not a hairstyle. A character is not an age, a job title, a wardrobe choice, or a list of surface attributes. "She is thirty-two, a marketing executive, wears her hair in a tight bun, drives a red convertible" — that is a police sketch. Visually identifiable and emotionally empty. A real character is defined by four things: what they want, what they fear, what they hide, and how they change by the end of the story. Everything else — the hairstyle, the car, the job — is delivery mechanism. The tight bun might signal control, but only if the story puts pressure on that control. The red convertible might signal freedom, but only if the character is running from something.

Without the engine of want, fear, secret, and transformation, your details are frosting on an empty plate.

This matters enormously when working with AI. The tools will enthusiastically generate surface detail. Ask an image generator for "a forty-seven-year-old man with weathered hands, a faded denim jacket, and a scar above his left eyebrow" and you will get back something that looks like a casting photo. But that man will not be a character until you know what he wants so badly he can barely admit it to himself, what he is terrified of losing, and what he is hiding from everyone — especially — himself.

I build every character from a six-line sheet. Not a forty-page backstory. Not a psychological profile. Six lines. Line one is the name — not just the name, but why that name. A character named Grace who has none carries irony. A character named Victor who has never won anything carries weight. Line two is the want: what they are actively pursuing, concrete and visible. "Wants to be happy" is useless. "Wants to sell the family restaurant before the bank forecloses" is a want that generates scenes. Line three is the fear: what they will do almost anything to avoid facing.

The best fears are specific and primal — not "fear of failure" but "fear of ending up like their father, who died broke and alone in the house he could not sell." Line four is the secret: what they are hiding from the other characters and usually from themselves. The secret creates subtext in every scene. It gives the actor — or the AI avatar — something to hold back, something that colors every interaction even when it is never mentioned. Line five is voice: how they speak. Not vocabulary level, but rhythm, sentence length, what they avoid saying, their default emotional register. Do they answer questions directly or deflect? Do they use humor as armor? Do they speak in fragments when emotional?

This line determines how you write their dialogue and how you describe their vocal delivery. Line six is change: how they are different at the end of the story versus the beginning. This is the arc in a single sentence. If you cannot write this line, you do not have a character arc yet. You have a situation with a person in it.

Here is what that looks like for the lead character, the protagonist of the sample short. Name: the lead character — forty-nine, a sound engineer who built a career mixing live concerts for mid-tier rock bands. The name is solid, unflashy, workmanlike. Like him. Want: To complete the posthumous album his wife the absent wife was recording when she died two years ago. He believes finishing it will honor her memory and give him closure. Fear: That moving on from grief means betraying the absent wife. That letting new people in means she is truly gone. That the best part of his life is permanently behind him. Secret: He has not listened to the raw tracks in six months. He tells everyone he is "working on the mix" but he opens the project file, stares at the waveforms, and closes it.

He is not healing. He is performing healing. Voice: Short sentences. Gravel from years of loud environments. Deflects emotional questions with practical answers. When cornered, goes silent rather than lie. Rarely says anyone's name directly — it creates too much intimacy. Change: He stops using the album as a shield against living and uses it as a bridge — finishing it not to close a chapter but to prove he can still create something that matters.

Those six lines give me everything I need to write the lead character. Every scene he is in, I know what he wants, what he is afraid of, what he is hiding, how he talks, and where he is headed. The backstory — the marriage, the illness, the funeral, the two years of isolation — all of it is implied by those six lines. I do not need to explain it. I need to show its effects.

If you are stuck on a character, write the secret first. Everything else flows backward from what someone is hiding. The want becomes what they pursue to avoid dealing with the secret. The fear becomes what happens if the secret gets out. The voice becomes how they talk when they are guarding something. The change becomes either releasing the secret or being destroyed by it. The secret is the key that unlocks the door.

Now, dialogue. This is where most AI-assisted screenplays fall apart, because AI models — particularly those trained with safety alignment — default to what I call "therapist speak." Characters articulate feelings with clinical precision. They validate each other's emotions. They use complete sentences and avoid interruptions. It sounds like a support group, not a conversation between human beings under pressure. Real people talk over each other. They trail off mid-sentence. They say the opposite of what they mean. They use slang, fragments, profanity when the situation calls for it. They lie, deflect, change the subject when uncomfortable. Your screenplay dialogue needs to sound like people talking, not like AI models being polite to each other.

What people actually say versus what AI thinks people say — the gap is real. AI thinks people say: "I understand you are hurting, and I want you to know that your feelings are valid." Actual people say: "You gonna stare at that all night or you gonna talk to me?" AI thinks people express grief by articulating it in complete paragraphs. Actual people express grief by going quiet, by snapping at someone they love, by reorganizing a kitchen drawer at 2 AM because sitting still is unbearable.

I fix this in two ways. First, I include dialogue style notes in every scene prompt. "the lead character speaks in short sentences when afraid. His voice uses technical language even discussing emotions. When cornered, he goes silent rather than lie." These behavioral anchors produce more distinctive speech patterns than any amount of generic "write realistic dialogue" instruction. Second, I read every line of dialogue aloud after generation. If it sounds like something an AI would write, I say it the way a real person would, then transcribe my version into the script. This manual pass is non-negotiable. AI gets you eighty percent of the way to good dialogue. The final twenty percent comes from human judgment about how people actually talk.

There is no button that replaces your ear.

The interruptions, the fragments, the subtext — these are not flaws in your dialogue. They are the features that make it human. A character who always finishes their sentences and never trails off is not eloquent. They are a robot wearing a mask. Let your characters step on each other's lines. Let them answer questions with questions. Let them say "I don't know" when they know exactly what they mean but cannot bear to say it. Subtext is the space between what is said and what is meant. It is where the audience lives.

All of this character and dialogue work feeds into a larger process — the AI writer's room workflow. I use a seven-stage system for every screenplay, and I do not skip steps. Professional screenwriters do not sit down and vomit out a hundred pages. They build in layers, locking each one before moving to the next. AI gives you speed of iteration, not permission to skip stages. You can generate a logline in thirty seconds instead of three days, but the sequence stays the same because the sequence works.

Stage one is the logline, and you lock this before anything else. Everything flows from it. If your logline is weak, your screenplay will be weak no matter how many drafts you run through AI. A strong logline for AI filmmaking needs to imply visuals — "a reclusive forger discovers his latest client is the assassin who killed his wife" implies faces, rooms, documents, tension. I can picture that film. AI can picture that film. Lock the logline by testing it: does it name a single protagonist? Does it contain an active goal and obstacle? Does it make someone want to know what happens? Do not proceed until the answer to all three is yes.

Stage two is the one-page synopsis — roughly four hundred to six hundred words that tell the complete story from beginning to end. This is where most writers reveal they do not actually have a story yet. They have a setup. The synopsis forces you to answer the hardest question: what happens in the second half? I write it as prose, not screenplay format. Act One sets up the world, protagonist, and inciting incident. Act Two traces rising action, complications, and midpoint reversal. Act Three delivers climax and resolution. The synopsis must include the ending. No ambiguity. You need to know how your story resolves because every scene either moves toward that ending or is dead weight.

Stage three is the fifteen-beat beat sheet. This is the structural skeleton — Opening Image through Final Image. The beat sheet runs two to three pages. When it is done, I should be able to read through all fifteen beats and feel the film's rhythm. If two consecutive beats feel similar in energy, I rewrite one. AI filmmaking suffers when scenes share the same emotional temperature. You need peaks and valleys. After generating the beat sheet, read only the odd-numbered beats aloud. These should feel like the film's escalating heartbeat. If beat eleven does not feel more intense than beat nine, your second half is flat. Fix it now. It is cheap to rewrite two sentences. It is expensive to rewrite twenty scenes.

Stage four is the scene-by-scene outline. The beat sheet tells you what needs to happen. The outline tells you where and when it happens, who is present, and what changes by the end of each scene. This is the document that separates writers from people who just talk about writing. It is necessary. It is tedious. I expand each beat into its constituent scenes. For each scene: scene number, location, time of day, characters present, one-sentence action summary, and critically — the change that occurs. If I cannot articulate what is different at the end versus the beginning, that scene does not exist. I use AI to generate the initial scene list, then review every entry myself. AI consistently proposes scenes that are too talky, too static, too interior.

Characters sit and think. I delete roughly a third of AI's suggestions and replace them with alternatives built around physical action. Movement gives AI video models something to render.

Stage five is writing actual scenes — one at a time. This is slower than asking for a full act, but the quality is dramatically better. AI loses coherence over long generations. One scene at a time, each anchored to its outline entry, produces consistent tone and focused action. I generate two to three versions of each scene and combine the best elements manually. Version A might have stronger dialogue. Version B might nail the ending. It takes minutes to combine them, and the result outperforms any single generation. Formatting matters. Have AI output in standard screenplay format. Clean formatting now saves hours later.

Stage six is the rewrite pass, and this is where I use a different AI model than the one that wrote the draft. This cross-model critique is one of the most effective techniques I have found. The second model reads the script cold, as a development executive would, and I ask specific questions: where is the motivation unclear? Which scenes repeat information? Where is the pacing slowest? If you had to cut ten pages, where would they come from? Each model has different strengths. Claude tends to be stronger on character psychology and thematic coherence. GPT tends to be sharper on structural pacing and commercial appeal. A critique from the model that wrote the draft will confirm its own mistakes rather than catching them.

The second model's ignorance of the generation process is an advantage — it cannot confirm the first model's assumptions. It can only respond to what is on the page.

Stage seven is the renderability review — the step most screenwriting books omit because they assume a human crew. AI filmmaking has hard constraints. A cinematographer can figure out a complex tracking shot through a crowd. An AI video model will produce visual gibberish. I go through the script scene by scene. How many characters on screen? Current AI handles one to three reliably. Four or more, and you get merging limbs and disappearing people. How complex is the action? A character walking across a room renders fine. Precise physical tasks — assembling machinery, playing an instrument with correct finger placement — will likely fail. How many locations? AI maintains consistency best in one defined space.

Rewriting for renderability is not a compromise of artistic vision. It is understanding your medium.

The voice distinction test is how I know my dialogue is working. I feed a different AI model all of one character's lines from the script, with no labels, no scene descriptions, no context — just the raw dialogue. I ask it to describe the speaking style and guess what kind of person would talk this way. Then I do the same for every other major character. If the model cannot distinguish the lead character from the absent wife, I have a voice problem. They sound like the same person with different names. The fix is usually in the six-line character sheet — I have not made their voices different enough at the foundation level. One character uses fragments when stressed. Another gets formal when angry. A third deflects emotional questions with jokes.

These differences need to be sharp enough that a stranger reading only the words could tell who is speaking.

When do you lock each stage? You lock it when changing it would require changing everything downstream. The logline locks when you cannot imagine a better one. The synopsis locks when someone who knows nothing about your project can describe the whole film from beginning to end after reading it. The beat sheet locks when the odd-numbered beats feel like an escalating heartbeat. The scene outline locks when every scene has a clear change statement. The screenplay locks when it passes the rewrite and renderability reviews. Locking does not mean the work is perfect. It means you are done searching for alternatives and ready to build on what you have.

Now, here is what I want you to think about, and I want you to really sit with this. Heavy dialogue films are the single most underexploited opportunity in AI filmmaking right now. Everyone is trying to make action movies. Everyone wants explosions and car chases and giant robots. The technology struggles with that. But two people talking in a room? The technology absolutely nails that. A face. A voice. A reaction. A pause. A look. This is what AI avatars were built for.

Think about it. A Sorkin-style walk-and-talk with avatars — rapid-fire dialogue, characters finishing each other's sentences, the camera pushing in as the stakes rise. An AI-age My Dinner with Andre — two people at a table for ninety minutes, and the audience cannot look away because the conversation is that compelling. A couples therapy session drama — the camera holding on faces as years of resentment surface in real time. A philosophical debate series — Nietzsche versus Kierkegaard rendered as a boxing match of ideas, with avatars in a darkened auditorium trading arguments like blows. Courtroom dramas. Job interviews gone wrong. Podcasts brought to life as visual conversations.

Twelve Angry Men in a jury room with AI-generated deliberation that shifts vote by vote.

The heavy dialogue genre is a goldmine that almost nobody in AI filmmaking is mining. They are all up the mountain with pickaxes looking for action sequences, and the gold is right here in the valley, in the conversations, in the rooms where people say things to each other that change their lives. Dialogue keeps production simple. Less movement means fewer technical problems.

One room means lower cost. Two or three characters means less chaos. The story moves through words, choices, secrets, jokes, arguments, and emotion. If the words are good and the performance is strong, the audience will stay. They will stay because people have been staying for great dialogue since the Greeks first put actors on a stage. The technology has changed.

The hunger has not.

Your next step is to build your characters, lock your logline, and run your story through the seven-stage workflow. Do not rush to the screenplay pages until the foundation can support them. When you finally write scene by scene, with locked outlines guiding every prompt, you will produce a draft that surprises you with its quality. Not because the AI is a genius. Because you were a good enough director to tell it what to do.

Chapter 7: The Lock — and the Freedom It Gives You

There's a moment in every film's life when the script stops being something you tinker with and becomes the blueprint you build from. In the old world, they made a big production out of it — the studio issued a "locked white" draft, the production office printed colored revision pages, and everyone on the crew knew that once those pages went out, changing a word cost money. Real money. Overtime money. Re-scout money. Re-build-the-set money. The lock was a budget document first and a creative document second, because in Hollywood the two things have always been the same piece of paper.

AI filmmaking doesn't have a production office. It doesn't have colored pages. What it has — and this is the part that will bite you harder than any budget overage — is the ability to change anything at any time for zero dollars. You can be three weeks into generating footage for your twelve-minute short, wake up at two in the morning convinced the third act needs to start eight pages earlier, and by dawn you have a completely rewritten scene feeding into your pipeline. The software doesn't care. The avatar doesn't push back. The voice clone doesn't mention overtime rates.

The barrier to changing your mind has never been lower, which is exactly why you need the discipline of the lock more than any studio executive sitting on fifty million dollars of other people's money.

I learned this the hard way, because I am a man who has historically needed to learn things the hard way. My first AI short was a six-minute thriller called The Interview, and I "locked" that script three separate times. The first lock held for about forty-eight hours before I decided the opening wasn't atmospheric enough. The second lock made it a week until I watched some video essay about nonlinear narratives and convinced myself my straightforward structure was "boring." The third lock actually stuck, but only because I'd already generated the opening sequence and could see, visually, that my rewrite impulse was garbage. Those two false locks cost me roughly twelve hours of production time that I'll never get back.

Twelve hours of my life, gone, because I couldn't tell the difference between "I think it's done" and "it is done."

Here's the distinction, and I want you to tattoo it somewhere visible. A script that is "pretty much done" is not done. It is Schrodinger's screenplay — simultaneously finished and unfinished, existing in both states until someone forces you to actually produce it. The cat is neither alive nor dead; the script is neither locked nor unlocked; and you, the filmmaker, are neither working nor finished, just suspended in a purgatory of perpetual revision that feels like productivity but is actually the opposite. When you lock, you make a decision: these words, in this order, tell the story I want to tell. I am now going to commit the time, the creative energy, and the rendering resources to bring them to life. Full stop. Period. Close the document.

The lock is a commitment device, which is a fancy psychology term for "a thing that makes it harder to change your mind." My wife uses commitment devices when she puts my blood pressure medication on the kitchen counter where I'll see it, because if it's in the cabinet I'll "get to it later" and later becomes never. A locked screenplay works the same way. It forces you to solve problems through execution rather than through the fantasy that some future version of the script will be so perfect that the problems disappear. They never disappear. You just get better at solving them in production instead of in revision.

In traditional film, a locked script functions as a budget document because every page costs money to shoot. In AI film, the locked script serves a different but equally vital purpose: it is a creative focus document. The constraint of the lock forces clarity. It tells your brain, which is already juggling avatar references and voice prompts and color palettes and music cues, that the story part is handled. The words are set. Move on. This matters because creative energy is not infinite. Every hour you spend rewriting scene four is an hour you're not spending making scene four exist. And at the end of the day, a rewritten scene that hasn't been generated is indistinguishable from no scene at all.

The hidden cost of post-lock changes is real even when the software feels free. Let me walk you through what happens when you make a "small" change after locking, using a concrete example. Say you're working on something like the sample short, and you've decided to move a story beat one scene earlier — the lead character finding a previously unknown recording the absent wife made in their old bedroom now happens in scene one instead of scene two. Seems minor. One little shift. But now your storyboards for scenes one and two are wrong. Every panel showing the lead character puttering around his home studio doing routine mixing work needs to be redrawn to show him frozen in his tracks, holding that DAT tape like it might bite him.

The quiet wide shots of the studio that you loved yesterday — the mixing boards glowing in the half-dark, vinyl records stacked on the shelf — no longer serve the right story beat. Four hours of storyboarding work, gone. Then the lead character's voice lines are out of sequence — the casual, muttered dialogue you generated for him tinkering with levels no longer makes sense, so you need new lines with that choked, discovery-heavy emotional reading. Add two hours. Then your avatar shots are partially invalid because the lead character's facial expressions don't match the new emotional content. New prompts, new renders, new review cycles. Three to four more hours.

Then the music cue you selected for the studio introduction — that gentle, ambient guitar piece — plays over a moment of raw discovery now instead of quiet atmosphere, and it's tonally wrong. Back to the music tool. Another hour. That single "small" change just consumed ten to twelve hours of production work, and that's assuming you caught the cascade immediately. In practice, you usually realize a script change has broken something three steps downstream, after you've already built on the old foundation. I've scrapped entire rendered sequences because a "minor" revision made them narratively incoherent with what came after. The software was free. My weekend was not.

This is why you need a filter. Not a casual "let me think about it" filter, but a real decision framework that you write down and follow like your doctor's sodium restriction. I use three questions, written on a sticky note that stays visible during every production phase. Question one: Is this change fixing an error, or is it an improvement? An error is something that doesn't work — a plot hole, a line that contradicts established character behavior, a scene that renders confusing rather than mysterious. An improvement is something that might be better — a sharper line, a more elegant transition, slightly faster pacing. You fix errors after locking. You do not implement improvements.

The time for improvements was during the review passes, which I'll get to in a moment. If it is not broken, you do not touch it. Question two: Does this change affect only this scene, or does it cascade? A local fix — correcting a single line of dialogue — is low-cost and sometimes necessary. A cascading fix — restructuring a scene, adding or removing a character beat — triggers the cost analysis I just described. You must write down every affected element before you start: storyboards, voice lines, avatar shots, music cues, sound effects. The full list. Only when you see the total scope do you decide whether the change is worth it. Question three, and this is the one most people skip: Will I still want this change after I've slept on it?

At least half of the post-lock changes I've been tempted to make felt urgent in the moment and obviously unnecessary the next morning. Institute a twenty-four-hour cooling-off period. Write the idea down, describe why you think it's needed, and do not act until the next day. If it still feels essential after sleep, proceed through questions one and two. If it feels less urgent — and it usually does — you just saved yourself hours of unnecessary rework.

Locking without reviewing is reckless, though. You don't lock the script because you're tired of working on it; you lock it because you've systematically verified that it's ready to become a film. I use four review passes, each with a specific mandate, and I do not rush them. The two to four hours you spend on these passes will save you twenty to forty hours of production rework, which is one of the better time investments you'll make in this entire process.

The Structure Pass is the big-picture review. You're not reading for word choice here. You're reading to confirm that the architecture of your story is sound — that every scene exists for a reason, that the sequence of events builds logically and emotionally, and that the pacing serves the story rather than your ego. For each scene, I answer three questions in writing, and I mean actually writing them down because thinking the answers in your head is how you lie to yourself. What does the protagonist want in this scene? Every scene needs a want — something the character is trying to achieve, learn, avoid, or understand. If you can't articulate a want, the scene is dead weight. Not atmospheric weight. Not thematic weight.

Dead weight that the audience will feel as a drag on momentum. Second question: What changes between the beginning and end of this scene? A scene where nothing changes is not a scene; it's a description of people standing in a room. The change can be small or large, but something must be different at the end than at the beginning. Third question, and this is the killer: What would be lost if this scene were removed? If you can remove a scene and the story still makes complete sense, the scene is optional. Optional scenes are luxuries that big-budget films can afford and AI filmmakers cannot. Every scene in your script should be load-bearing — remove it and the structure weakens or collapses.

The Dialogue Pass is where you read every spoken line out loud. Not in your head. Out loud, with your actual voice, in your actual room, probably sounding like a crazy person to anyone listening through the wall. This feels ridiculous until you do it and immediately discover three problems you never noticed while reading silently. Dialogue has exactly one job in a screenplay: to reveal character while advancing the story. Not to be clever. Not to show off your vocabulary. Not to deliver exposition disguised as conversation. Character and story advancement, simultaneously. Every line that does only one of these — or worse, does neither — needs to be rewritten or cut.

The most common dialogue problem I see in AI filmmaking scripts is what I call the "explainer monologue." Because AI voice tools can generate long, perfectly enunciated passages without actors getting tired, there's a temptation to let characters speak in paragraphs instead of sentences. A human actor would push back on a five-line uninterrupted speech because they know it sounds unnatural. Your AI voice tool will render it beautifully and destroy the scene's credibility in the process. Read each line and ask: would this character say this, in this way, at this moment? Not "could they say this" — anyone could say anything. Would they?

A grieving sound engineer who's been alone in his studio for three days doesn't deliver perfectly structured philosophical observations. He mutters, he trails off, he talks to himself, he plays the same four bars over and over while trying to figure out what his dead wife was trying to tell him. Your dialogue should sound like it was overheard at a bus station, not composed at a desk. One practical technique: after you've read the dialogue out loud, try covering the character names and seeing if you can tell who's speaking from the words alone. If every character sounds interchangeable, you have a writer-voice problem, not a character-voice problem. Fix it before you lock.

The Renderability Pass is the one that doesn't exist in traditional screenwriting manuals because it doesn't need to. If a traditional screenwriter writes "the camera pans across a crowd of ten thousand," the production team figures out how to shoot it. In AI filmmaking, if you write that same scene and your image generation tool struggles with crowd consistency, you have a problem that will derail your production. Scene by scene, you ask: can the AI tools I plan to use actually produce this? Not theoretically. Not "I saw someone do something similar on Reddit." Can your specific stack deliver this visual and audio content at the quality level your film requires? How many characters are on screen at once?

Most AI tools struggle with multiple characters in the same frame — consistency drops, faces distort, limb counts become unreliable. What's the action level? A character sitting at a desk talking renders very differently from a character running through a forest being chased. What are the setting and period requirements? A contemporary home studio interior is easy because the training data is full of them. A 1940s wartime field hospital requires period-accurate details that the AI may get subtly wrong. During the renderability pass on my second AI film, I discovered that a crucial scene — a character driving while having an emotional phone conversation — was going to be a nightmare.

AI video tools struggle with vehicle interiors, with the relationship between driver and steering wheel, with phone-holding gestures. I could have burned days trying to force it to work. Instead, I rewrote the scene: the character is now pulled over, sitting in a parked car, talking through the windshield. The emotional content was identical. The production feasibility went from nightmare to straightforward. That is the point of this pass — to find those problems while you're still working with words, not rendered frames.

After the three passes, I apply one final filter that I call the Cut or Combine Rule, and I've stolen it from painful experience. Every scene must change the story in a way the audience can perceive. Not "changes the story for me because I know the backstory." Changes the story in a way that an audience member watching for the first time will register and care about. The test is simple: write one sentence describing what is different at the end of the scene compared to the beginning. Then ask: would the audience notice if this difference did not exist? If the answer is no, you have two options — cut the scene entirely, or combine it with an adjacent scene so the change happens within a scene that does matter.

This is where many writers falter because it requires killing material you like. You'll have scenes with dialogue you're proud of, visual descriptions that evoke exactly the mood you wanted, character moments that feel authentic. The rule does not care. If it does not change the story, it is a luxury, and luxuries are for filmmakers with infinite render budgets and no deadline. You are not that filmmaker. Cut it or combine it, lock the result, and move forward.

The lock discipline I'm describing here — these passes, these questions, this commitment — it's not just about preventing chaos. It's about enabling a kind of filmmaking that barely existed before. Because when you can lock a script quickly and confidently, you unlock a rhythm of production that changes everything.

Think about what rapid lock-to-ship cycles make possible. the lead character is a filmmaker I know through an online forum — never met him in person, but I've watched his work for two years. He locks a new five-minute film on the first of every month. Every month. By the fifteenth he's generating footage. By the twenty-fifth he's in sound mix. By month-end he's exporting and publishing. His films are small — single locations, two characters, tight constraints — but they're finished. Twelve films a year. At that pace, he's not just building a portfolio; he's building an audience that knows to come back. His viewers have learned his release rhythm the way people used to learn television schedules. The lock is what makes that possible.

Without it, he'd be on draft six of his first film, still "improving" it while the lead character from an alternate universe already shipped his fourth.

Or consider serialized content with what I call the mini-lock. A weekly series where each episode gets its own small lock ceremony — less drama than a feature lock, but just as real. Episode one locks on Monday, ships by Sunday. Episode two locks the following Monday. Each episode has its own production lists, its own extraction documents, its own identity. The weekly rhythm builds muscle. You stop treating every script like it might be your last and start treating it like what it is: one episode in an ongoing conversation with your audience. The mini-lock is less intimidating than the feature lock, which means you actually do it instead of postponing it indefinitely while you chase some fantasy of perfection.

Then there's what I call the living film. This one excites me more than almost anything else in AI filmmaking. You lock version one. You ship it. Your audience watches, reacts, comments, asks questions. And instead of moving on to a completely new project, you lock version two — incorporating what you learned from that first audience response. Not everything they suggest. You are the filmmaker, not a polling service. But maybe a scene that confused people gets clarified. Maybe a character who landed well gets more depth. Maybe the ending that half your audience loved and half hated becomes a branching choice in version two. The lock doesn't prevent evolution. It creates a checkpoint. Version 1.0 is locked forever.

Version 1.1 gets its own lock ceremony. The film becomes a living document, versioned like software, each lock a milestone in an ongoing creative dialogue. This is impossible in traditional film — imagine telling a studio you're going to "patch" the theatrical release. In AI filmmaking, it's just another Tuesday.

What the Lock Really Unlocks

Here's where I want you to stop thinking about process and start thinking about possibility — because rapid locking discipline doesn't just make you more efficient. It opens doors that most filmmakers don't even know exist.

Think about this: if you can lock, produce, and ship a five-minute film in thirty days, you can respond to current events while they're still current. Not the news cycle — I mean cultural moments, emerging conversations, ideas that are alive right now. A traditional documentary takes years. Your AI documentary-short on a breaking story? Locked in a week, shipped while people still care. What happens when your audience realizes you're the filmmaker who always has something relevant to say?

Or consider the experiment. In traditional film, trying a wild visual style or an unconventional narrative structure is a massive gamble — months of production, thousands in budget, all riding on one creative hunch. With rapid lock discipline, you can ship a bold experimental short in two weeks, learn whether it works, and apply those lessons to your next "regular" film. Failure becomes cheap and fast instead of expensive and slow. How many creative risks would you take if failure cost you two weeks instead of two years?

The mini-lock enables something even stranger and more wonderful: serialized stories that adapt to their audience without losing artistic integrity. Imagine a twelve-episode mystery where episode four reveals a twist that nobody predicted — including you, the filmmaker, who locked episode one without knowing exactly how the threads would resolve. Each mini-lock lets you adjust, tighten, build on what worked. The audience feels the story breathing because it actually is. You're not improvising — you're iterating within a disciplined framework. That's a distinction with a difference.

And the living film? That's where it gets genuinely strange in the best possible way. Version one ships. Your audience debates it. Version two ships with the debated scene extended. Version three ships with an alternate ending unlocked by viewer choice data. You're not making a film anymore — you're cultivating a film, tending it like a garden, letting it grow branches. What does it mean to "direct" something that keeps changing? What does it mean to "release" something that never really ends? I don't have definitive answers. But I know this: the filmmakers who figure out what living films can become will be defining a new art form, not just adapting an old one.

The lock is what makes all of this possible. Not the technology — the discipline. The ability to say "this version is done" so you can ship it, learn from it, and make the next one better. Without that commitment, you just have infinite drafts. With it, you have a practice, a rhythm, a body of work that compounds on itself. That's not a workflow optimization. That's a creative superpower.

The lock discipline applies far beyond narrative films, and this is where a lot of people miss the point entirely. A documentary filmmaker locks her shot list — the specific archival clips she needs, the interview questions she's settled on, the locations she's confirmed. A branded video producer locks the client-approved script and the visual reference pack before a single frame gets generated, because the client sign-off IS the lock. An educator building a twenty-module online course locks each module's script independently, treating every lesson as its own small film with its own extraction lists and its own voice line inventory. The lock is not a narrative filmmaking tool. It's a commitment tool.

It works anywhere someone needs to stop planning and start producing.

And here's the psychological shift that I think matters most. In Hollywood, the lock was a budget document. It existed to prevent money from leaking out of the production. The assumption underneath it was: change is expensive, so minimize change. In AI filmmaking, the lock becomes a creative focus document. It exists to protect your attention, not your wallet. The assumption underneath it is: your creative energy is finite, and every hour you spend rewriting is an hour you're not spending making something visible. This changes the entire psychology of production. You're not locking because some accountant in a corner office is nervous about the quarterly report.

You're locking because you respect your own time and energy enough to commit them to execution. The lock becomes an act of creative self-respect. That's a very different feeling than signing off on a budget spreadsheet. That feeling — the feeling of being someone who commits, who executes, who ships — that's what separates people who talk about films from people who actually make them.

Now, here's where I want you to think about something bigger than the process. The lock changes your relationship with your work in a way that I find almost spiritual, and I don't use that word lightly. When you lock a script, you shift from being someone who might make a film to someone who is making a film. The difference is everything. And the beautiful thing is that the lock concept scales to whatever you're making. A thirty-second branded video for a client? You can lock that in an afternoon — the script is basically a voiceover and three visual beats. A twenty-minute short film? Give yourself a week of review passes, because the complexity compounds. An episodic series?

You need a season bible lock first — the overarching story, the character arcs that span episodes, the visual rules of your world — before you lock individual episode scripts. What about interactive content, where the viewer chooses paths? Some paths lock because they're central to the narrative spine; others branch because the whole point is exploration. Documentary-style content where the "script" is really a shot list? You lock the shot list, the interview questions, the archival footage inventory — whatever constitutes your production blueprint. The lock adapts to whatever you're making because it's not about the format. It's about the decision to commit.

Different genres require different lock disciplines, too, and this is worth understanding before you start. Heavy dialogue films — two-handers, stage adaptations, courtroom dramas, Sorkin-style ensemble pieces — are the sweet spot for AI avatar production because dialogue is what the tools do best right now. Your lock needs to be airtight on the words because the words carry the entire film. Scenes requiring heavy special effects integration — a character hallucinating, a sci-fi battle sequence, a dream sequence — need an additional layer of renderability verification before lock because if the AI can't generate the visual, your locked script describes something impossible.

The point is: know your genre, know your tools, and calibrate your lock discipline accordingly.

Once the four passes are complete and you've made your final revisions, you transition from writer to producer. The locked script becomes a source document, and your job is to extract from it every piece of information your production pipeline will need. I create four extraction lists from a complete reading of the numbered script. The Character List records every speaking character with their name, approximate age, physical description, emotional disposition, and which scenes they appear in. For AI filmmaking, the physical description needs extra detail because it becomes part of your image generation prompts.

"the lead character, late 40s, unshaven, weary eyes, rumpled flannel shirts, moves through his home studio like a man who hasn't slept well in months" is a usable prompt component. "the lead character, the protagonist" is not. If a character appears only in recordings or flashbacks, note that too — the absent wife's descriptions will emphasize the difference between her recorded presence and the present-day reality the lead character inhabits. The Location List maps every setting with its priority level — hero locations that anchor multiple scenes, secondary locations that appear briefly, tertiary locations that flash by in a single shot.

The Prop List inventories objects that appear on screen, with special attention to anything involving text or mechanical detail because AI generation often garbles those. The Sound Needs List captures every audio element beyond dialogue — music cues, ambient atmosphere, specific sound effects, Foley moments. These four lists become the foundation of your production phase.

Scene numbering happens before any of this extraction work. Number sequentially from the first scene to the last using whole numbers — scene 1, scene 2, scene 3. The scene number goes into every production file name associated with that scene. Storyboard panels become "SC03_panel_01." Voice lines become "SC05_the lead character_line_03." Rendered avatar shots become "SC07_avatar_closeup_v02." This naming convention seems obvious until you're three weeks into production with two hundred files and no systematic way to find anything. Format the locked script as a clean, single document. Remove all revision marks, comments, and alternative versions.

Save it with the date in the filename: "Father_Frequency_LOCKED_2026-03-15." Every time you make an authorized post-lock change — after passing the three-question test — update the date and append a brief change log at the end describing what changed and why.

The final step is mechanical but psychologically crucial. Export the locked script and save it in at least three places: your local working drive, a cloud backup, and ideally an additional offline location. I email the locked PDF to myself with the subject line "FATHER FREQUENCY — LOCKED IN PROGRESS — [date]." This creates a timestamped record independent of my file management habits. Then — and this is the part that requires actual discipline — you stop editing the script file. Close it. The working document is now a reference document. If a genuine error emerges during production, you make the fix in the production files directly and document the change in your log. You do not reopen the script and start rewriting.

The script is a contract between your past self, who wrote the story, and your present self, who is producing it. Both parties benefit from honoring the agreement.

The freeze is a milestone worth acknowledging. You have done something significant: you wrote a complete screenplay, reviewed it systematically, and committed to turning it into a film. Most people never reach this point. They start scripts, they revise endlessly, they chase perfection into oblivion. You locked. You are producing. The next phase — visualization and storyboarding — takes your words and begins the transformation into something the audience can see and hear. And it all starts with the simple, terrifying, liberating act of saying: these words, in this order, are the ones.

Chapter 8: Seeing the Film Before It Exists

Your screenplay is locked. Every scene earned its place, passed the dialogue test, survived the renderability check, and got cut or combined until the whole thing moves like a story should. Now comes the transition that separates people who finish films from people who accumulate very expensive folders of disconnected pretty pictures. You are moving from words to images. This is where most AI filmmakers stumble, and they stumble for a reason that has nothing to do with technical skill and everything to do with process.

They open their image generation tool, type a prompt that feels right for the first shot they can picture, generate something beautiful, and then try to figure out what comes next. Thirty renders later they have thirty gorgeous stills that don't cut together, don't share a visual language, and don't add up to a watchable scene — let alone a clear and connected film. I've seen it dozens of times. I've done it myself. The temptation is nearly irresistible because generation is fun, and planning feels like homework, and most of us are still kids inside who want to skip to the good part.

The antidote is storyboarding. Not the elaborate hand-drawn panels you see in making-of documentaries for big studio films. Not professional illustrator renderings that cost thousands of dollars per sequence. A working storyboard — fast, functional, specific enough to keep every subsequent decision aligned with every previous decision. Your storyboard is the bridge between the screenplay you locked and the rendered footage you will soon generate. It answers the question that no prompt, however well-written, can answer on its own: what does this film actually look like, shot by shot, from the first frame to the last?

Here's the thing that surprises people: AI filmmakers need storyboards even more than traditional filmmakers do. A traditional filmmaker on location has physical constraints that enforce a kind of accidental coherence. The sun is where the sun is. The location looks like it looks. The actors are wearing the costumes they're wearing. If you shoot six shots in the same room on the same day, those shots will share a lighting environment, a spatial geometry, and a color reality whether you planned them or not. The universe provides continuity for free. AI generation offers no such gift. Every prompt is a blank slate.

You can generate a kitchen scene in warm golden light at noon, then generate the reverse angle in cool fluorescent light at midnight, and the tool will give you both without complaint because the tool doesn't know what "the same kitchen" means. It only knows what you tell it, and what you tell it changes with every prompt. This means the AI filmmaker faces a continuity challenge that traditional filmmakers simply don't have. You are not preserving reality across shots. You are constructing reality, shot by shot, from noise and latent space, and if you don't plan that construction, the seams will show through every frame.

The cost of not storyboarding follows a pattern I've seen repeat so predictably I could set my watch by it. First comes style inconsistency. Your opening scene looks cinematic and desaturated because you were influenced by the reference images you happened to have open that morning. Your third scene looks bright and oversaturated because you saw a beautiful sunset photo and decided to chase that feeling. Your seventh scene goes dark and high-contrast because you were in a different mood. Viewed individually, each scene might contain beautiful images. Cut together, they feel like three different films. The audience senses the disconnection and disengages, even if they can't articulate why something feels off.

Then comes missing coverage. Without storyboarding, AI filmmakers generate the shots they can picture easily and skip the shots that are harder to visualize. You end up with six beautiful close-ups of your protagonist's face and no wide shot establishing where he is. You have three stunning images of the radio equipment but no insert of his hand reaching for the dial that would bridge the conversation to the tension beat. The edit becomes a nightmare of gaps.

Then come the expensive re-renders. Two weeks in, seventy shots deep, you realize the color temperature you established in Shot 1 doesn't match what evolved organically in the scenes you generated later. The opening feels disconnected from the rest of the film, so you go back and regenerate it, and possibly the next three shots, and possibly the establishing shots in scenes four and six. One uncaught decision early in the process cascades into a dozen correction renders that eat your budget and your schedule.

Every minute you spend planning your visuals before touching a generation tool saves you an hour of re-rendering, style-matching, and frustration on the back end. I know that sounds like an exaggeration but it isn't. I've tracked it across projects. The ratio holds.

Aspect Ratio, Palette, and Style

Before you create a single storyboard panel, you must decide your aspect ratio, and you must never change it. This is the most consequential visual decision you will make, and it's the easiest to get wrong because it feels trivial. It is not. Your aspect ratio determines how your audience's eye moves across the frame. A 2.39:1 ultrawide ratio creates horizontal scanning — the eye sweeps left to right, taking in vistas and spatial relationships between characters separated by distance. It feels cinematic because it mimics the way human peripheral vision works in wide open spaces. A 1.85:1 or 16:9 ratio feels more natural, more "real," because it approximates the field of view of human attention.

A 4:3 ratio forces vertical composition, stacking elements in ways that create claustrophobia or nostalgic association. A 9:16 vertical ratio — increasingly relevant for mobile-first distribution — flips the entire compositional logic, emphasizing height, falling, rising, vertical stacks of information. Once you choose your ratio, every single panel, every single generation, every single video render must use it. Write it on a Post-it note and tape it to your monitor. Mixed aspect ratios signal either a creative choice so bold it needs justification, or a filmmaker who didn't plan. You know which category you're in right now.

Then comes your color palette — three to five recurring colors that become your film's signature. This doesn't mean every image contains only those colors. It means those colors are present often enough, and in positions of enough visual importance, that they become the film's color fingerprint. A viewer seeing any single frame should be able to guess, with better than random odds, that it comes from your film and not someone else's. Your palette serves a practical function beyond aesthetics. When you include your film's color keywords in your generation prompts, you want those keywords to be consistent. If your film's palette is steel blue, burnt orange, and beige, you include those words or close synonyms in every prompt.

This gives the AI model a repeated signal that gradually biases its output toward your intended look. Changing your color keywords from shot to shot produces the visual equivalent of a film shot by three different cinematographers who never met.

I remember an early project through NJ Film Productions where we had a clear vision — moody interiors, warm practical lighting, shallow depth of field. The first batch of images came back gorgeous. Then I took a weekend off, returned with what I thought was the same prompt, and the images looked like a different crew had shot them. The warmth had shifted toward orange. The shadows had gone from rich brown to cold blue-gray. I lost three days because I hadn't locked down my visual parameters in writing. Don't be me. Choose your palette, write it down, enforce it.

Style references come next — images that communicate your film's look. You need two kinds. General style references communicate the overall look of your film and go at the front of your storyboard document or on a dedicated mood board page. Specific shot references show the exact composition, angle, or lighting approach you want for a particular panel, and these sit next to the panels they inform. When you generate a storyboard panel using AI, your prompt should include both a description of the specific shot and a reference to the general style. "A suburban street at dawn, cool blue tones, one window glowing warm orange, wide shot, Todd Hido photography style, cinematic, 2.39:1 aspect ratio" tells the model both what to show and how to show it.

The reference artist name activates visual associations in the model's training data that bias the output toward the qualities you want. Without that reference, you get a generic suburban street. With it, you get a suburban street that carries the specific emotional weight of blue-hour isolation.

The Anatomy of a Storyboard Panel

Each storyboard panel needs eight elements, and every panel you create should have all eight even if some are implied rather than explicitly written. The shot type — wide, medium, close-up, insert. The camera angle — high, low, eye level, Dutch. The lens feel — wide angle stretching space, telephoto compressing layers, normal approximating human vision. The subject placement — centered for stability, rule-of-thirds for dynamic tension, edge placement for isolation. The lighting description — what's doing the illuminating and how. The mood — a single word like "lonely" or "uneasy" that captures the emotional register. The color palette reference — which of your chosen colors dominate this frame.

And a unique identifier — scene number, shot number, version letter — so you can connect the panel to its prompt and its rendered output without going insane three weeks from now.

I want you to think about what this process unlocks, because most AI filmmakers are making people talking in rooms, and that's fine — people talking in rooms is the foundation of drama — but the tools can do so much more. Storyboarding for AI film is actually more liberating than traditional film because you can visualize impossible shots and the AI can generate them. A camera moving through a keyhole. An interior view of a human bloodstream. The surface of Mars at sunset. A bird's-eye view of a city where the buildings are made of glass and the streets are water. In traditional film, these shots require enormous resources or are literally impossible. In AI film, they require only that you plan them.

The storyboard is where you plan. A sci-fi film set entirely on a generation ship — all interiors, all controllable through carefully constructed location packs — becomes achievable for a single creator. A noir detective story in perpetual rain where the atmospheric conditions are consistent because you locked them in your style bible. A fantasy courtroom drama in a world where gravity works differently. A period piece where the period is actually consistent because you did the research and built the reference library. These aren't distant fantasies. They're current capabilities that almost nobody is utilizing because almost nobody plans deeply enough.

From Storyboard to Shot List

From the storyboard, you build your shot list — the document that tells you how to make each shot, when to make it, and what it needs to match. The panel shows you what the shot looks like. The shot list tells you the practical execution details. For each panel, verify that it contains all eight anatomy elements. Check that the shot logically connects to the shots before and after it — does the direction of action, the eyeline, the screen direction make sense? A character exiting frame right in Shot 4 should enter frame left in Shot 5 if it's a reverse angle. Note what each shot needs to match — the same character's appearance, the same location's lighting, the same color temperature.

Without these matching notes, you will generate Shot 7 in isolation and create a discontinuity that breaks the viewer's immersion.

Grouping shots for efficient rendering is one of those techniques that sounds like an unnecessary optimization until you try it and realize it changes everything. Group by location first — render all the kitchen scenes together so you stay in the same visual zone. Group by emotional register second — I don't render a desperate, high-tension confrontation and then immediately try to render a quiet, contemplative insert. The emotional whiplash affects the quality of my prompt writing. Batch all the tense scenes together when I'm in a tense creative headspace, and all the quiet scenes together when I'm calmer. Within each group, render in script order so you can verify continuity as you go.

If Shot 2 doesn't match Shot 1, you catch it immediately rather than discovering the mismatch two days later after you've rendered twenty other shots.

Location Packs: Your Visual Bible

Now, location packs. This is where AI filmmaking gets really interesting, because the single most underappreciated advantage of these tools is location access. Not speed. Not cost. Access. I can put characters inside the Oval Office, on the bridge of a starship, in a medieval throne room, or on the observation deck of a skyscraper that doesn't exist yet — all before lunch, without permits, without insurance, without explaining myself to a location manager who wants to know if my crew of twelve will fit through the service entrance.

For the sample short, I generated the entire location pack from my desk. the lead character's home studio — an intimate workshop space where he mixes recordings and confronts his grief. The wide shot showing the cramped room with equipment stacked on every surface. The close-ups of the mixing board with its rows of knobs and faders. Shelves of vinyl records lining the back wall. The window with its slowly changing afternoon light, the one bit of the outside world that seeps into the lead character's isolation. The doorway where his wife the absent wife used to stand, watching him work, now just an empty frame. Every image in the pack was generated from the same shared reference phrase, the same seed numbers, the same material palette.

All of it created without setting foot outside my own door.

But unlimited access creates a problem: inconsistency. When everything is possible, the temptation is to generate every shot independently, treating each frame as its own mini-masterpiece. The mansion looks one way in the establishing shot, subtly different in the interior dialogue scene, and completely wrong in the climactic confrontation. Viewers notice this even when they can't articulate it. Visual consistency is one of the deep cues that tells an audience "this is a professional production." When environments shift between cuts, the brain registers disruption. The viewer trying to figure out why the kitchen changed between shots is not engaged with your story.

A location pack is your defense. It's a pre-generated set of six to twelve images showing the same fictional space from multiple angles, under different lighting conditions, with different compositional framings. Think of it as your location's visual bible — the reference set that every subsequent shot must conform to.

I build each pack methodically, starting with the wide establishing view that shows the full room, its proportions, and major architectural details. Then the opposite angle from the reverse corner — essential for dialogue scenes and over-the-shoulder framing. Then detail views of distinctive architectural elements, key props, textural surfaces. The entry and exit view from the primary doorway. An alternate lighting version — day, night, and a dramatic variant for tense scenes. An environmental context shot — through a window for interiors, the immediate surroundings for exteriors. Simpler locations need six images; complex hero locations need ten to twelve with additional detail views and overhead perspectives.

Build sequentially — wide first, then opposite angle, then details, checking each new image against the established set before moving on. One mismatched image early in the process will propagate inconsistency through everything that follows. Use a shared reference phrase — a distinctive cluster of descriptive words copied into every prompt for that location, functioning as its fingerprint. "Large room with furniture" is too generic. "Brutalist concrete bunker with recessed LED strip lighting" is a signature. Combine that with seed numbers from your image generator and image references uploaded alongside your text prompts, and you have a three-layer consistency system that keeps your locations coherent across dozens of shots.

Five categories of visual information must stay locked across every image in a location pack. Architecture and spatial layout — wall positions, ceiling height, doorway locations, room proportions. Include a simple verbal floor plan in every prompt: "Large rectangular room, twenty by thirty feet. Fireplace centered on the north wall between two tall windows." Props and set dressing — fixed furniture, artwork, decorative objects that help viewers subconsciously confirm they're in the same space. Lighting quality and direction — windows stay in fixed positions, lamps stay on the same walls, even across different times of day. Period details — every element must belong to the correct historical moment or fictional design language.

And material palette — limit yourself to five or six primary materials used across every location. A dystopian future might use raw concrete, rusted steel, cracked safety glass, faded yellow paint. Recurring materials unify the world in ways that viewers feel before they can name.

Building Your Location Inventory

Before generating a single location pack, audit your script and create a complete location inventory. Classify each location by priority. Hero locations anchor multiple scenes and carry significant screen time — they get the full twelve-image treatment with multiple lighting variations, detailed prop lists, dedicated seed numbers.

the lead character's home studio in the sample short appears in seven scenes. It gets the hero treatment — twelve images covering every angle, three lighting variations for the different times of day when the lead character works, detailed prop lists for the mixing board, the vinyl collection, the coffee mug that never leaves his desk. That's our primary world, and it needs to feel lived-in.

The one secondary location in the sample short — a small cemetery where the lead character visits the absent wife's grave in two scenes — gets a medium pack of six to eight views. The iron gate at the entrance, the row of headstones, the particular bench where the lead character sits, the tree that casts shadows across the grass in late afternoon. It's a simple location, but it still needs the same consistency discipline. Same gate, same tree, same light quality every time we cut to it. Nothing about a smaller location means you get sloppy.

Tertiary locations appear once or briefly and get minimal packs of four to six views generated quickly with basic consistency measures. This inventory prevents you from discovering halfway through production that you forgot to design a location that appears in three critical scenes, or from wasting time building elaborate packs for locations that flash by in a single two-second shot.

The naming convention matters more than you think. I use: PROJECT_LOCATION_VIEW_LIGHTING_VERSION.ext. "FF_HomeStudio_Wide_Day_v02.png" tells me everything I need to know without opening the file. Store packs in folders organized by priority — Hero, Secondary, Tertiary — and keep a master spreadsheet tracking location name, priority, views generated, seed numbers, shared reference phrase, palette, prop list, and connected locations. The ten minutes you spend on organization will save hours of hunting later.

World-Building for Genre and Period Films

World-building for genre films takes everything I've described and raises the stakes. Science fiction and fantasy present the greatest consistency challenge because there's no real-world reference for audiences to validate. When you show a Victorian parlor, viewers have seen hundreds of them and their brains know what to expect. When you show a starship engineering deck, their only standard is internal coherence — does this world obey its own rules?

Coherence breaks down predictably. The bridge has circular displays, but engineering shows rectangular screens with different interface logic. The throne room is black volcanic stone, but the royal bedchamber is pale marble with no explanation. These inconsistencies read as carelessness. The solution is a design language document — written visual rules governing your fictional world. Define your material palette, your technology level and design logic, your scale and proportion rules. Generate your most important location first, the one that establishes your world's visual identity most clearly, then use that location as an image reference for all subsequent locations.

Create one "signature technology" visual element that appears in every location — a specific light fixture style, a recurring geometric pattern — that functions like a logo for your world.

Period pieces are easier than sci-fi because you have reference material — photographs, paintings, surviving architecture — and harder because audiences notice anachronisms instantly. Research your period thoroughly before generating. Collect references from authentic sources. Build a period-appropriate material list and treat it like a shopping list: "1920s Manhattan apartment: hardwood floors, area rugs, plaster walls, heavy drapery, brass fixtures, dark wood furniture, radio console, hallway telephone." Pay obsessive attention to lighting sources. Before 1880, night interiors are dramatically dark — candles, oil lamps, fireplaces produce limited, warm illumination with deep shadows.

By the 1920s, electric lighting appears in affluent homes but with different color temperature than modern LEDs. The AI knows "old-timey" vaguely but not that plate glass windows weren't common in England before the 1680s. You must provide the specificity, or the AI defaults to generic approximation.

If you're starting your first AI film project, consider setting it in the present day in a location you've actually visited. This isn't a creative limitation — it's a strategic advantage. You know what a modern coffee shop looks like because you sat in one this month. The AI's training data favors contemporary imagery. And anachronism isn't a risk — a smartphone in a character's hand is appropriate, a flatscreen TV is appropriate. Every period detail you don't have to police is energy you can invest in story and craft. As you gain experience, expand gradually. Your second project might be set in the 1980s. Your third might venture into science fiction, applying the consistency discipline you've built to a world without real-world reference.

The filmmaker who masters consistency in a contemporary setting will build better fantasy worlds than the one who tries to invent everything at once.

The Power of Planning

I want to close this chapter with a thought about what all this planning makes possible. Most AI filmmakers are making small things — short scenes, explainer videos, social content — and there's nothing wrong with that. But the location pack concept, the style bible, the systematic storyboarding process — these are the same tools used to build worlds in major films, and they're available to you now, at your desk, for the cost of a subscription and the discipline to plan.

A generation ship where every corridor connects logically because you built the location packs before generating a single shot. A noir city where it rains in every exterior shot because you locked the atmospheric conditions in your style bible. A fantasy kingdom where the architecture, the technology, the materials all feel like they belong to the same civilization because you wrote the design language document first. These aren't distant dreams. They're current capabilities that almost nobody is utilizing to their full extent. The tools are there. The question is whether you'll do the planning work to use them well.

Your storyboard is your film in miniature before the miniature becomes real. Your location packs give your fictional spaces the weight and memory of real places. Your style bible keeps every frame speaking the same visual language. Build these things with care, and the generation phase becomes execution rather than exploration. You will know what you're making because you already decided. The only thing left is to make it.

SECTION 2: PRODUCTION

Scaling, Pipelines, and the Future

The reader enters this section with a finished first film and asks: what now? By the end, they know how to turn one project into a sustainable practice — series, features, anthologies, and a workflow that improves with every film.

Chapter 9: What Project Two Feels Like

I finished exporting the sample short at 11:47 PM on a Tuesday. The render took forever because I'd added too many effects in DaVinci Resolve, which is what you do when you don't know what you're doing and hope that complexity equals quality. I watched the final file play through once, nodded at the screen like a lunatic talking to himself, and went to bed without telling my wife it was done. I needed a night to sit with it.

The next morning, I came downstairs and opened my laptop, and something hit me that I hadn't expected. I looked at my desktop and saw a complete studio. Not metaphorically — literally. The HeyGen tab was still open with the lead character's Digital Twin loaded. ElevenLabs had my voice clone settings saved with all the tags I'd spent weeks refining. My project folder had subfolders for audio, B-roll, dialogue renders, music stems, color references. I had prompt documents full of camera descriptions that worked and lighting setups that didn't. I had a Seedance account with credit history. I had a DaVinci Resolve project template with my preferred timeline settings already configured.

I'd accidentally built a production company. And I was ready to make another film without having to figure any of that out again.

That's the moment I want you to have. Project one is learning. Project two is where efficiency kicks in. Everything you built, broke, fixed, and optimized for your first film becomes the foundation for everything that comes after. The question isn't whether you can make a second film. The question is how much faster and better you can make it because the hard infrastructure work is already done.

Your Template Library: The Cookbook Approach

After the sample short, I started a document I call my “Recipe Book.” It’s embarrassingly simple — just a text file organized by category — but it saves me hours on every new project. Whenever I write a prompt that produces a great result, I copy it in. Not the vague memory of what I wrote. The exact prompt, word for word, with a note about what it produced and what I changed to get there.

My camera descriptions section has about forty entries now. “Medium close-up, 85mm lens, shallow depth of field, warm interior lighting” — that one produces reliably cinematic faces. “Wide establishing shot, dolly-in movement, morning fog, desaturated color palette” — that’s my go-to for opening B-roll. I don’t have to reinvent these from scratch anymore. I pull from my cookbook, adapt to the specific scene, and move on.

The lighting section is even more useful. I wasted three days on the sample short trying to get a nighttime interior look that felt moody without being muddy. I tried maybe twenty variations before landing on one that worked. That winning prompt is now in my book, tagged “night interior, practical light sources, film noir influence.” Next time I need that look, I’m starting from what works, not from zero. The failures live in there too, with notes about why they failed. “Too much ‘cinematic’ in the prompt produces generic results.” “Adding specific color temperatures helps consistency.” These notes are gold. I paid for them with frustration.

Voice direction tags live in their own section. I keep the exact ElevenLabs settings that produced the lead character’s grief-stricken delivery — the stability score, the style exaggeration, the specific combination of tags that gave his voice that hollow quality in the scene where he finds the absent wife’s last recording. I also keep notes on what didn’t work, because those memories fade and you don’t want to relearn expensive lessons. “Whisper tags below 0.3 stability produce artifacts.” “Adding ‘resigned’ to the emotion list worked better than ‘sad’ for the final monologue.” This level of specificity is what turns a hobby into a craft.

Asset Reuse: Your Digital Repertory Company

Here’s something that blew my mind when I realized it. the lead character’s Digital Twin — the one I built in HeyGen using my own reference photos and refined over weeks of adjustments — can play a different character entirely. Different wardrobe prompt, different hairstyle, maybe glasses. The underlying face model is the same, but the presentation changes enough that audiences won’t recognize him unless they’re looking for it.

I tested this for my next project after the sample short. I put the lead character’s Twin in a blue work shirt with a different lighting setup, gave him a more energetic voice setting in

ElevenLabs, and suddenly he was a completely different person. My wife watched the test clip and didn't recognize the base model at all. "He looks like a guy who fixes things," she said. Which was exactly what I was going for.

the second sample short used this approach. I didn't need to build a whole new Twin from scratch for Harold Vance — I adapted an existing one with wardrobe changes and voice adjustments. Saved me probably a week of setup time. Harold became a distinct character because of his performance, not because of his underlying geometry. That's what actors do, after all. They bring different energies to different roles while looking fundamentally the same. De Niro has played a boxer, a taxi driver, and a mobster with the same face. The performance is what makes them different people.

Your home studio environment can become multiple locations with the right prompts. The room where I filmed the lead character's close-ups became, with different lighting descriptions and set dressing prompts, three different interiors across two projects. Your music library expands with each film. Your sound effect collection grows. You're not starting fresh each time. You're building a studio inventory, the same way a traditional production company keeps a props room and a costume closet. Only your props are digital and your costume closet is a text document full of wardrobe prompts.

The Iteration Mindset Applied to Workflow

After every project, I do a post-mortem. Not a fancy one — I drink coffee and write down three questions in a notebook. What took too long? What broke? What surprised me by working better than expected?

For the sample short, the voice work went into the "surprised me" category. The ElevenLabs PVC training for the lead character's voice was a revelation — once I had the training data right, the emotional range I could get from text prompts was far beyond what I'd expected. That discovery led me to invest more heavily in voice performance for the second sample short, pushing the vocal characterization of Harold Vance harder than I would have otherwise. I gave Harold a specific vocal rhythm — slightly formal, a touch old-fashioned, the way a man who spends his days with dead people might sound — and it worked because I'd learned from the lead character what these tools were capable of.

The first round of B-roll for the sample short was a disaster. I generated maybe thirty location shots before realizing they all had a same-y quality — that smooth, slightly unreal look that marks AI-generated environments. I threw most of them away and started over with more specific, more unusual prompts. That lesson went straight into my workflow for project two: write weirder prompts for B-roll, push the descriptions into specificity instead of generic beauty. "Abandoned gas station at dusk, graffiti on the pumps, one fluorescent light still buzzing" produces something far more interesting than "beautiful cinematic establishing shot." Specificity is the difference between a location and a backdrop.

The color grade took too long because I was learning DaVinci Resolve as I went. By project two, I'd taken a free online course and built a basic understanding of nodes and color wheels. I also created a starting LUT — a custom color preset — that gets me 70% of the way to my preferred look with a single click. I refine from there, but I'm not building the foundation every time. Each project should leave you with one reusable improvement. Not ten. One. Over five projects, that's five solid improvements to a workflow that was already functional.

Scheduling: The Calendar Is a Budget Too

A short film takes four to eight weeks from script to export, and I don't care what anyone on YouTube tells you about making one in a weekend. Could you generate something in a weekend? Sure. A finished, coherent, emotionally satisfying short film? Not unless you've already done ten of them and have your pipeline running like a Swiss watch. And even then, the weekend claim is suspect.

My four-to-eight-week breakdown looks like this. Week one is script and pre-production — finalizing the story, building any new assets I need, scheduling my voice recording sessions (which means finding two hours when my house is quiet and I can perform without my wife wondering why I'm shouting about grief in the basement). Weeks two through five are production — generating footage, building scenes, recording and refining dialogue, pulling together music and sound elements. Weeks six through eight are post-production — editing, color, sound design, mixing, that final pass where you watch the whole thing five times in a row looking for anything that pulls you out.

A feature, which I haven't made yet but am actively planning, will take six to twelve months. That's not padding. That's the reality of sustaining a coherent story across ninety minutes, managing dozens of scenes, keeping visual consistency, and not burning out halfway through. The calendar is a budget just like your wallet is a budget. Underestimate it and you'll end up with half a film and a lot of regret. I've seen it happen to friends who got excited, started shooting, and ran out of steam because they didn't plan for the long middle of the process where nothing seems to be coming together.

The sample short's Post-Mortem: The Honest Version

Looking back at the sample short with the clarity of distance, here's what actually happened. The voice work was the strongest element — not because I'm brilliant, but because I spent real time on it. I recorded myself reading the lead character's lines dozens of times, trying different deliveries, keeping the takes that had genuine emotion and discarding the ones that felt performative. The ElevenLabs PVC clone preserved that work beautifully. When the lead character whispers the absent wife's name in the third scene, that catch in his throat is my throat. I know because I remember recording it at 10 PM, thinking about my own losses, and getting it right on the sixth try.

The sound design was the other win. Every recording the lead character plays in that film — the static, the distant voices, the fragments of the absent wife's speech — was built by hand in Audacity. Layering noise textures, EQing until the voices felt like memories rather than recordings, adding just enough distortion to suggest degradation without making it unpleasant to listen to. That work is 100% human. No AI generated those sounds. I want that said plainly, because it's easy for critics to claim AI filmmaking has no craft, and they're wrong. The hours I spent getting that static to feel like grief made physical — that's craft. No algorithm suggested those frequencies. I found them by listening, rejecting, adjusting, listening again.

What went poorly? The first round of B-roll, as I mentioned, was unusable. I'd written prompts that were too generic, too aspirational, too Pinterest-board-in-sentence-form. "Beautiful cinematic shot of city at night." Meaningless. Every AI generator has seen a million variations of that prompt and will give you the average of all of them. I learned to write prompts like I was directing a cinematographer with a specific vision. Low angle, wet asphalt reflecting sodium vapor streetlights, a single figure in the distance, anamorphic lens flare. Specificity is what separates usable footage from wallpaper.

The color grade took forever because I didn't know what I was doing. I'd open DaVinci Resolve, stare at the color wheels, move things around until it looked different, and hope it looked better. By the end of project two, I'd learned to start with a reference — a still from a film whose look I wanted to approximate — and work toward matching it. Much faster. Much more intentional. Roger Deakins shoots every movie with a look in mind. I learned to do the same, digitally.

the second sample short: Ten Days of Applied Learning

I made the second sample short specifically to test whether my lessons from the sample short had stuck. Four minutes. One main character. Three locations, all interiors. A simple premise — Harold Vance discovers his dead mentor has left him a series of increasingly inappropriate pranks to execute at the funeral home — that let me focus on performance and timing rather than world-building.

the second sample short cost about thirty dollars. Most of that was a month of a music subscription I used to find the right dark comedy score. I already had the Twin. I already had the workflow. I already knew which prompts worked and which didn't. I wrote the script on a Sunday, generated Harold's footage across three evenings, spent a weekend on sound design — the whoopee cushion had to sound exactly right, and let me tell you, there's an art to recording a whoopee cushion that reads as funny rather than just sad — and edited the whole thing the following week.

It worked. Not just technically — the film got into two festivals and audiences laughed in the right places. But it proved that my pipeline was real. I could make films faster and better because I'd done the hard learning work once and documented it. The limerick eulogy scene, where Harold reads a dirty poem at a funeral while his black suit gets progressively more disheveled, and the final coffin recording prank — both landed exactly the way I heard them in my head. That never happened on my first film. On my first film, almost nothing played the way I heard it in my head.

Pro Tip: After every project, write a "lessons learned" document. Not in your head. On paper or in a file. Two paragraphs minimum. What worked, what didn't, what you'd do differently. I keep mine in the same folder as the project files, labeled "POSTMORTEM_READ_THIS.txt." Six months later, when you're starting something new and you've forgotten half your hard-won wisdom, this document will save you. I open the the sample short post-mortem before starting every new project. It's become a ritual, like checking your car's oil before a long drive.

Pro Tip: Build your prompt library like a cookbook — recipe on one side, result on the other. I use a simple format: the exact prompt, followed by a description of what it produced, followed by any adjustments I made. "Downtown street at dawn, wet pavement, neon signs reflecting" produced good establishing shots but the neon was always too bright. Added "neon signs muted, overcast sky" and that fixed it. That progression is valuable. It's not just the winning prompt that matters. It's knowing how I got there.

Pro Tip: Keep a casting directory. I have a document with every Digital Twin and Photo Avatar I've built, with notes. "the lead character base model — works for 40s-50s male, neutral to sad emotions, not great for action or comedy." "Harold variant — older male, excellent for deadpan comedy, timing works best with longer pauses." This saves me from reinventing characters. When a new project needs a certain type, I know who in my digital repertory company can play the role. I'm up to six distinct characters now, and I add notes every time I learn something new about one of them.

Common Mistake: Starting project two before reflecting on project one. I get it. You're excited. The tools are right there, humming, ready to go. You want to ride the momentum. But if you don't stop and catalog what you learned, you're going to make the same mistakes again. I did this between my first and second attempts at the sample short — rushed into a new version without understanding why the first version failed, and wasted two weeks generating footage I didn't use. Reflection is work. Do it anyway.

Common Mistake: Trying to make project two bigger instead of better. This is ego talking. Your first film was small, so your second film needs to be epic, with more characters, more locations, a more complex story. No. Your second film needs to be better — tighter, more confident, more emotionally precise. the second sample short was smaller than the sample short in scope and budget, but it was more successful because I knew what I was doing. Scale comes later. Competence comes first.

Common Mistake: Throwing away your project one assets instead of organizing them. I see people delete their working files after export to save hard drive space. Don't do this. Storage is cheap. Your project folder — properly organized with clear subfolder names and a README file explaining what's what — is the foundation of your studio. Compress it if you need to, back it up to cloud storage, but don't discard it. You'll want those assets again. the lead character may show up in another film someday, and when he does, I'll have everything I need to bring him back.

Common Mistake: Not documenting what you learned because you think you'll remember. You won't. I'm sixty-five and I can't remember what I had for breakfast half the time. Write it down. Future you will be grateful, and future you is the person doing all the actual work from now on. The best filmmakers I know are obsessive documentarians of their own process. Not because they think anyone else will read it. Because they know their own minds are unreliable narrators.

What If Your Pipeline Became Something Bigger?

Here's where I let myself dream a little, and I think you should too.

What if your "project two" wasn't one project at all? What if it was ten — a series of micro-films, two to five minutes each, released weekly on a platform that actually lets you build an audience? With your pipeline established, each one becomes faster to produce. The first takes two weeks. The fifth takes one. By the tenth, you're making a complete film in five days. A weekly series of cinematic shorts, each standing alone but building a body of work that no individual piece could achieve on its own. That's not a fantasy. That's a schedule. I've been thinking about this seriously. A series called "Short Circuits" — little emotional shockers, three minutes each, released every Friday. I could do it. The pipeline would handle it.

What if you built a shared asset library with other filmmakers you trust? I know three other AI filmmakers in various parts of the country. We're talking about sharing our prompt libraries, our music collections, our sound design elements. Not the finished films — those are ours — but the raw materials. If ten filmmakers each contribute a hundred prompts, everyone gets a thousand. The community approach that built open-source software could build filmmaking resources just as effectively. I think about the hours I spent getting the lead character's voice right. Those settings could help someone else make their own character's voice work. Why keep that knowledge private?

And here's the one that really keeps me up at night in the excited way. What if your Digital Twin became a repertory actor? The same base model, playing different roles across a dozen films — a grieving husband in one, a comedic funeral director in another, a noir detective in the third, a documentary narrator in the fourth. Each performance distinct because of how you direct the voice, how you style the appearance, how you frame the shots. Your own personal actor, available whenever inspiration strikes, who knows your methods and responds to your direction because you built them. That sounds like science fiction. But I already have two distinct characters living in my HeyGen account, and I'm working on a third.

By this time next year, I could have a repertory company of six or eight.

The pipeline you build for project two is the studio you're going to run for the next decade. Treat it like the asset it is. Organize it. Document it. Improve it with every project. The films you make are what the world sees, but the pipeline behind them is what makes those films possible. That's your real business. That's NJ Film Productions — not the individual projects, but the system that produces them reliably, efficiently, and with room to grow.

My laptop is open. the lead character is waiting. Harold is waiting. The next character — whoever they turn out to be — is already half-formed in my head, and I know exactly which buttons to press to bring them to life. That's not a small thing. That's a studio.

Time to make the next one.

Chapter 10: Scaling Up — Series, Anthologies, and Features

I was sitting in my kitchen at eleven o'clock at night, staring at the final render of the sample short, when my wife walked in and asked the question that changed everything. She'd seen the rough cuts, heard me curse at DaVinci Resolve more times than she could count, watched me obsess over whether the lead character's navy sweater looked consistent from scene to scene. She knew this film inside and out without ever sitting down to watch it straight through. And now, with the finished file playing on my laptop, she watched the last thirty seconds — the shot where the child character puts on her mother's headphones and closes her eyes, hearing the absent wife's voice one more time — and she turned to me and said, "So what happens next?"

I thought she meant the plot. I started explaining how I wanted the ending to be ambiguous, how the audience should decide for themselves whether the child character finds peace or just more pain. My wife shook her head. "No, Jim. I mean what happens to the lead character? What happens to a supporting character? What happens to the child character?" She pointed at the screen. "You made me care about these people. You can't just leave them there."

I opened my mouth to argue. They're not real, I wanted to say. They're Digital Twins. They're pixels and voice clones and a script I wrote in three weeks. But she was right. I did care about them. And the tools that made the sample short — HeyGen for the characters, ElevenLabs for the absent wife's voice, Seedance for the atmosphere, DaVinci Resolve for the edit — those tools were still sitting on my computer. the lead character's Digital Twin didn't disappear when I exported the final file. His voice model was still there. His face was still there. The apartment set was still there. I'd spent six weeks building a world, and I'd only used it for twelve minutes.

That's when it hit me. the sample short didn't have to be a film. It could be an episode.

The Epiphany of Reuse

Think about it like this. Chuck the plumber once told me about a job he did in an office building downtown. The building manager called him in to fix a toilet on the third floor. Chuck fixed it in twenty minutes. Then the manager said, "While you're here, the one on five has been running." Then six. Then the basement. Each one took less time because Chuck had his tools out, he knew the building's pipe layout, he was already in rhythm. By the end of the day he'd fixed eight toilets and the building manager paid him a flat rate that made them both happy. Chuck said it was his most profitable day that month, and the individual toilets got cheaper with every one he did.

That's your episodic model right there. Episode one of anything is expensive because you're building everything from scratch. You're training your Digital Twins. You're designing your locations. You're figuring out the visual language, the color grade, the pacing. You're learning the tools while you use them. But episode two? Episode two is Chuck walking up to the fifth floor. The tools are already out. The world is already built. You just need a new story.

I started sketching out what the sample short Episode Two might look like. Same characters. Same apartment, though maybe a supporting character's living room this time. A new crisis — maybe the lead character finds a recording he made with the absent wife that he'd forgotten about, something that changes how he remembers her. Same pipeline. Same voice models. Same aesthetic. I estimated the budget in my head and laughed out loud. Episode one cost me $847 and six weeks. Episode two would cost maybe $300 and two weeks. Most of that cost was just compute time and my own coffee.

The economics are almost funny if you've ever worked in traditional production. I once visited a friend on the set of a streaming series in Brooklyn. They had eighty people on set that day. Trailers lined three blocks. The craft services budget alone was more than my entire film. And here's me, sitting in my kitchen, realizing I could produce six episodes of a series for less than the cost of one day of their parking. Not the same quality level, sure. Not the same scale. But the same emotional truth. The same ability to make an audience care.

Anthologies: Many Worlds, One Pipeline

Neighbor Frank came over one afternoon while I was working on this idea. He saw my whiteboard covered in notes — the sample short episodes on the left, something called the second sample short expanded into a three-part miniseries on the right, and in the middle a list of other concepts I hadn't even started yet. Frank squinted at it like he was reading a foreign language. "Jim," he said, "you're gonna make all of these?"

"Maybe," I said. "Or maybe I make one of each and see what sticks."

That's the anthology model. You build a pipeline that can handle anything, and then you run different stories through it. The tools don't care whether your film is a thriller about a grieving sound engineer or a dark comedy about a funeral director with a whoopee cushion problem. HeyGen doesn't know the genre. Seedance just generates what you tell it to generate. DaVinci Resolve edits grief and comedy with the same timeline tools. The pipeline is the pipeline.

An anthology lets you test. You make three short films — twelve minutes each, say — and you release them together. They share a sensibility, maybe a visual style, maybe a theme. Loss and what we do with it. That's what connects the sample short and the second sample short if you think about it. Harold Vance loses his mentor and inherits pranks. the lead character loses his wife and inherits her voice. Two completely different tones, same human territory. An anthology called something like "What We Leave Behind" could hold both stories and three more besides.

And here's what nobody tells you: an anthology looks like a studio made it. Seriously. If you release five short films that share a visual quality level, that have consistent production values, that feel like they came from the same creative mind — which they did, yours — distributors and festival programmers treat you differently. You're not a guy with a short film anymore. You're a filmmaker with a body of work. That distinction matters more than you think.

The Feature-Length Leap

Now let's talk about the elephant in the room. You're making twelve-minute shorts. You're thinking about episodes and anthologies. Eventually some voice in your head — maybe yours, maybe your mother's — is going to ask the big question: "When are you going to make a REAL movie?" By which they mean a feature. Ninety minutes. The kind of thing that plays in theaters or at least gets listed on IMDb with a respectable runtime.

Here's the truth. The tools don't change when you go from twelve minutes to ninety. Not really. HeyGen generates the same characters whether the scene is thirty seconds or three minutes. Seedance doesn't care about runtime. ElevenLabs voices the absent wife's lines the same way in minute eighty as it did in minute eight. The AI tools scale almost infinitely.

What changes is the management. And I can't emphasize this enough: making a feature is not ninety minutes of filmmaking. It's ninety minutes of decisions, multiplied by a hundred.

Let me break this down practically. the sample short had, what, fourteen scenes? Maybe sixteen if you count the transitions. I could keep the whole thing in my head. I knew what happened in scene seven because I'd just finished scene six. the lead character's sweater was consistent because I generated all his shots in two days and I remembered what I'd done.

A ninety-minute feature might have a hundred and twenty scenes. Maybe more. You're not going to keep that in your head. You're going to open your project file on week eight and stare at a timeline that looks like a piece of abstract art, and you're going to think, "Did I already generate the scene where a supporting character visits the apartment? Or was that a different scene? Did the lead character have the elbow patches in this scene or did I generate it before I added those?"

This is where organization becomes everything. And I mean boring, methodical, Chuck-the-plumber-would-be-proud organization.

You need a folder structure before you generate your first shot. Not after. Before. Something like: Project Name > Characters > the lead character > Outfits > Navy Sweater, Gray Henley, Funeral Shirt. Project Name > Locations > Apartment > Living Room, Bedroom, Kitchen. Project Name > Scenes > Scene_01_Opening, Scene_02_Headphones, and so on. Every generated asset gets named with a convention: Character_Location_Outfit_Versions. the lead character_Apartment_Sweater_v03. If that sounds tedious, it is. If that sounds unnecessary, wait until you're thirty scenes into a feature and you can't find the right version of a shot.

You need a show bible. A simple document — Google Docs is fine, Notion if you're fancy — that tracks every creative decision you've made. What color is a supporting character's cardigan? Green. What shade of green? Write it down. What earrings is she wearing? Silver hoops. Write it down. What does the lead character's voice sound like at different emotional registers? Document it with reference clips. Every time you make a creative choice, you put it in the bible. Because six months into production, you will not remember. Nobody remembers. That's not a failure of memory, that's just being human.

Grumpy Old Man Tool Note: Google Docs and Notion

Google Docs and Notion are where you keep the thinking organized. A film has too many moving parts to trust your memory, especially if your memory is like mine and occasionally forgets why you walked into the kitchen. Use a living document or workspace to track characters, rules, locations, prompts, voice settings, style choices, legal notes, and what you already decided three Tuesdays ago.

The possibility is a show bible that talks to your whole AI pipeline. Your character descriptions can feed Midjourney. Your scene notes can feed Seedance. Your voice rules can guide ElevenLabs. Your continuity notes can help Claude evaluate the script. The document becomes the brain of the project, and the tools become the hands.

The limitation is over-organization. Some people spend more time decorating their Notion dashboard than making the film. That is not productivity. That is scrapbooking for anxious adults. Keep it simple. If the system helps you create, use it. If it becomes another hobby, close the tab and make a scene.

When to Bring in Humans

I need to say something that might sound like it contradicts the whole premise of this book. The one-person studio has limits. I believe almost everything I've written in these pages. But I'm not going to pretend that a single person can do literally everything at feature scale. You can't. Or rather, you can, but the result won't be as good.

Sound is where you'll hit the wall first. AI-generated dialogue — whether it's text-to-speech from ElevenLabs or cloned voices — works beautifully for most scenes. But a feature has scenes with overlapping dialogue, with background noise that needs to be carefully layered, with subtle audio cues that build atmosphere over ninety minutes. At some point you need a re-recording mixer. Not a full sound department. One person with good ears and a decent setup who can take your AI-generated stems and mix them into something that feels cinematic.

Color grading is the second place. DaVinci Resolve's free version is miraculous. I've said that a dozen times in this book and I'll keep saying it. But a feature-length color grade is a different beast than a twelve-minute short. If you can learn to do it yourself, great. If you can't, a freelance colorist for two or three weeks is worth every penny. And it's not that many pennies compared to traditional post-production budgets.

Music is the third. AI-generated music from tools like Udio or Suno has gotten shockingly good. I've used it. I'll keep using it. But a feature needs score that evolves, that has themes and variations, that knows when to be silent and when to swell. If you're not a composer — and I'm not — consider finding one. Not a full orchestral score. A single musician with a laptop and some talent can do things that AI music tools still struggle with. I know a guy in Newark who scores indie films for three thousand dollars. That's not a typo. Three grand and he gives you something that transforms your film.

So when do you hire humans? When the cost of not hiring them is higher than the cost of hiring them. That's it. That's the rule. If bringing in a re-recording mixer for a week costs eight hundred dollars and the difference between a self-mixed feature and a professionally mixed feature is the difference between getting distribution and not getting distribution, that's not a hard math problem. Do the math.

The Economics Don't Make Sense (In Your Favor)

Let me put some numbers on the table because I know you're wondering. A traditional independent feature — ninety minutes, modest ambitions, no name actors — costs between five hundred thousand and two million dollars. That's the real range. I've been to the film markets. I've seen the budgets. Half a million is scraping by. Two million is comfortable but still indie.

An AI-assisted feature, made with the workflow I've described in this book, costs between five thousand and fifteen thousand dollars. I know that sounds insane. I know it sounds like I'm lying or leaving something out. I'm not. Here's how that breaks down.

Software subscriptions for a year: maybe two thousand if you're subscribing to the premium tiers of everything. Compute credits for generation: three to five thousand depending on how complex your visuals are. The Digital Twins you train, the Seedance generations, the voice work in ElevenLabs. It adds up but it doesn't explode. Music licensing or original score: one to three thousand. Festival submissions and marketing materials: one to two thousand. Contingency: a thousand. That's your ten to fifteen thousand right there.

At the low end, if you're careful, if you use free tiers where possible, if you do your own sound mix and color grade? Five thousand. Maybe less. I made the sample short for $847. Scale that up by a factor of six for runtime, add some complexity, and you're still under ten grand.

Now here's the kicker. A fifteen-thousand-dollar AI feature that looks like it cost two hundred thousand dollars is worth two hundred thousand dollars to a distributor. Not fifteen. Two hundred. That gap — the arbitrage between what you spent and what it looks like you spent — that's your advantage. That's why the traditional industry is terrified of this moment. Not because AI filmmaking is going to replace Hollywood. Because a filmmaker in New Jersey with a laptop just closed the budget gap by a factor of fifty.

Pro Tip: Block Production

If you're planning a series — three episodes, six episodes, whatever — don't make them one at a time. Block your production. Spend one intense week generating all your shared assets for the entire season. Every outfit for every character. Every location from every angle. Every establishing shot. Do it all at once while you're in the zone, while you remember the look you want, while the tools are behaving.

Then spend the next week writing and refining your scripts. Then spend subsequent weeks generating episode-specific scenes. But that first block — the shared assets — that's your foundation. Get it right and everything after is faster, cheaper, and more consistent. I learned this the hard way on the sample short. I generated the lead character's shots sporadically over six weeks, and there are two scenes where his sweater looks slightly different because HeyGen's model drifted between sessions. For a series, block production eliminates that problem entirely.

Common Mistake: The Feature Before the Short

I see this in online forums all the time. Somebody discovers AI filmmaking, watches a few tutorials, and announces they're making a feature. Their first film. They've never completed anything longer than a TikTok video and now they're planning ninety minutes of cinematic art.

Don't do this. Please. I'm begging you as someone who has made bad films and learned from them. Your first AI film will teach you things no tutorial can. Your second film will be exponentially better because you'll apply those lessons. Your third film is where you start to know what you're doing. A feature is not three shorts played back-to-back. It is its own discipline. If you haven't made a short, you don't know what you don't know. And what you don't know will destroy your feature.

Make the sample short first. Make the second sample short. Make something twelve minutes long that you finish, that you screen, that you show to people who aren't your relatives. Then make another one. Then, maybe, start thinking about something bigger. The tools will still be there. They'll be better, actually. There's no rush. The only person pushing you to make a feature before you're ready is your own ego. Tell your ego to wait in the car.

What If You Went Wild With This?

Here's what keeps me up at night — in a good way. What if you committed to a weekly AI-generated series for a year? Fifty-two episodes. Twelve minutes each. Ten and a half hours of content. That's more runtime than most streaming series get in two seasons.

The cost would be maybe fifteen thousand dollars total. Less than one episode of a traditional streaming show costs per minute. You'd release weekly on YouTube or your own website. You'd build an audience that comes back every week. After a year, you'd have more content than most indie filmmakers produce in a decade. You'd be one of the most prolific storytellers working. And you'd have done it from your kitchen table.

Or what if your anthology became a channel? Not a film, a destination. Viewers come for one short film and discover ten more. You curate other AI filmmakers' work. You become the platform. I've seen this happen with podcasts. I've seen it happen with YouTube channels. I haven't seen it happen yet with AI film anthologies, which means the opportunity is still there.

Or the feature. The real one. The ninety-minute film that plays at a festival and gets picked up and finds its audience. Made by you. Written by you. Directed by you. Produced by you. For less than the cost of a used car. That's not science fiction. That's next year. That's someone reading this book right now, someone who's going to finish their short, learn their craft, and then make the feature that changes their life.

The pipeline scales. The tools scale. The only question is whether you'll scale with them. My money says you will. Now let's talk about where all this is heading, because the future is coming faster than you think.

Chapter 11: Where This Is All Heading

I remember the exact moment I saw my first truly terrible AI-generated video. It was January of 2023, and a friend sent me a link with the message "Have you seen this?" I clicked it and watched thirty seconds of what was supposed to be a woman walking through a park. Her hands melted into each other like candle wax. Her legs phased through the bench she was supposedly sitting on. The physics looked like a fever dream designed by someone who had never seen a human body move. I laughed. I showed my wife. We both laughed. Then I tried it myself and made something even worse. A man — theoretically — trying to drink coffee. His hand went through his face. The coffee cup merged with the table.

It looked like a glitch in the simulation, which I suppose it was.

I closed the program and didn't open it again for three months. I figured this was a toy, a party trick, something for Reddit threads and tech demos. It wasn't filmmaking. It wasn't even close.

Then in June, someone posted another video. Same tool, supposedly. A woman walking through a park. But this time her hands were — I couldn't believe it — actual hands. Five fingers. Moving naturally. The bench was a bench. She sat on it. The physics worked. It wasn't perfect. A trained eye could spot the tells. But it was five, maybe ten times better than what I'd seen in January. In six months. Not six years. Six months.

That's when I understood. The technology only improves. It never gets worse. That melted-hand video from January? That's the worst AI video that will ever be made. Every day after that, the floor got higher. The ceiling too. And the rate of improvement wasn't linear — it was accelerating. What took six months in 2023 took six weeks in 2024. By late 2024, the improvements were coming weekly. You'd wake up on a Tuesday and a tool you used on Monday had a new feature that changed what was possible.

I thought about that a lot while making the sample short. Every limitation I bumped into — a hand position that looked slightly off, a background that wasn't quite consistent between shots — I reminded myself that this was the worst the tool would ever be. Next month it would be better. Next year it would be unrecognizable. The question wasn't whether the technology would get good enough. The question was whether I would get good enough along with it.

The Trajectory Is Your Friend

Here's what the curve looks like if you plot it honestly. Video quality from AI generation has improved roughly tenfold in the past two years. Character consistency — the ability to make the same face appear in different shots, different angles, different lighting — has gone from basically impossible to routine. Voice cloning has gone from robotic and uncanny to indistinguishable from human speech in most cases. Lip sync, camera movement, environmental detail, emotional expression, fabric simulation, hair physics — every axis you can measure has moved in one direction only.

And it's not slowing down. The companies building these tools — Runway, HeyGen, Luma, ElevenLabs, Kling, and the two dozen startups you've never heard of that will be household names next year — they're competing in the most productive way possible. Each new release forces the others to match or exceed it. It's a race, and the prize is nothing less than the future of visual media. They're pouring billions of dollars and the brightest engineers on the planet into making these tools better, faster, cheaper. This isn't like the dot-com bubble where companies burned cash on nonsense. These tools generate revenue. People pay to use them. That means the companies survive and keep improving.

What does this mean for you, practically? It means that a film you make today will look dated in two years. That's not a reason to wait — it's a reason to start now. Because the film you make today teaches you the craft that you'll apply with better tools tomorrow. the sample short made with 2025 tools looks good. the sample short remade with 2027 tools will look like it cost fifty thousand dollars instead of eight hundred. The story won't change. the lead character and the absent wife and the child character will still be the same people. But the pixels will be better. The hands will be perfect. The lighting will be indistinguishable from a real apartment at magic hour.

I'm not telling you to remake your films every two years. I'm telling you that your next film will automatically be better than your last one because the tools will be better. You get to focus on story, on performance, on the things that don't change, while the technology handles the things that do.

What's Coming (The Honest Forecast)

I don't have a crystal ball. Neighbor Frank bought me one as a joke for my seventieth birthday and all it did was sit on my desk and collect dust. But I can read the tea leaves. I pay attention to what the research labs publish and what the product teams ship. You can draw a straight line from where we are to where we're going, and the line points up.

In the next one to two years, here's what I expect with high confidence. Hands will be solved. Completely. The six-finger nightmare is already mostly gone, and within a year it'll be a story you tell new filmmakers to make them feel better about their own struggles. Longer continuous shots — right now most AI video is limited to a few seconds of coherent motion before things drift. That limit will extend to thirty seconds, then a minute, then multiple minutes. Character rendering across sessions will get so consistent that you'll be able to generate scene forty of your feature and have the lead character look exactly like he did in scene three, without heroic effort on your part.

Real-time generation — seeing your scene take shape as you adjust prompts — will cut your iteration time by a factor of ten. You'll go from ten generations to get the right shot to one or two.

I also expect we'll get proper multi-character scenes. Right now, getting two characters to interact naturally in the same frame is hard. The tools treat each character separately and the results can feel like two people pasted into the same shot. Within two years, that problem will be largely solved. Characters will look at each other properly. They'll hand each other objects. They'll stand at consistent distances. When a supporting character hands the lead character the box of the absent wife's tapes in the sample short, it'll look as natural as if two actors had actually performed the scene.

In the two-to-five-year range, things get more speculative but still grounded in what we can see developing now. AI-assisted directing is coming. Not AI replacing the director — I don't think that's what most of us want — but AI reading your script and suggesting shot compositions. You write "the lead character sits alone in the dark, listening to the absent wife's recording," and the system proposes three different framings: a wide shot emphasizing the empty room, a close-up on the headphones, an over-the-shoulder on the recorder. You choose. You direct. The AI assists. That's a tool I'd pay for today if it existed.

AI editing — rough assembly from footage — is closer than people think. The technology to watch raw scenes and identify the best takes, the natural cut points, the emotional beats, already exists in prototype at several companies. In three years I expect to load my generated scenes into an AI editor and get back a coherent rough cut in minutes. Not a finished cut. Not something I'd release. But a starting point that saves me ten or fifteen hours of assembly work, which I can then spend on the fine cutting where human judgment actually matters.

AI sound design that understands emotional context is the one I'm most excited about. Right now I manually select every sound effect, every ambient layer, every musical cue. I have to think about what a grieving apartment sounds like at two in the morning and then find or create those sounds. In a few years, I think you'll describe the emotional register of a scene — "tense, lonely, anticipating something that won't come" — and the system will generate a soundscape that matches. You'll adjust it, refine it, make it yours. But the starting point will be miles ahead of silence, and it'll understand that the lead character's apartment sounds different at the beginning of the sample short than it does at the end because grief changes how you hear everything.

The Industry Response (Some Will, Some Won't)

Let me tell you what's going to happen in the traditional film industry because I've watched this movie before, so to speak. Some established filmmakers will embrace these tools immediately. They'll see AI generation as another tool in their kit, like Steadicam or digital color correction. They'll use it for pre-visualization, for effects shots, for sequences that would be impossibly expensive any other way. These filmmakers will thrive because they'll be able to realize visions that budgets previously prevented.

Others will fight it tooth and nail. They'll call it cheating. They'll say it isn't real filmmaking. They'll form advocacy groups to exclude AI-generated work from festivals and awards. Some of their concerns will be legitimate — copyright questions need real answers, labor protections for people who work in effects departments are genuinely important. But some of their resistance will be about protecting their position. When you've spent twenty years climbing a ladder, you don't love the idea of someone building an elevator.

New distribution models will emerge. We're already seeing it. Film festivals are creating AI-specific categories. Streaming platforms are acquiring AI-generated content because it's good and it's affordable. Audiences — this is the part the critics miss — audiences don't care how a film was made. They care if it moved them. If it did, they'll watch another one. If it didn't, they'll scroll to the next thing. The production method is invisible to the experience of watching.

Audience expectations will shift in subtle ways. As AI filmmaking gets better, viewers will become accustomed to seeing production values that used to require studio budgets. The indie film that looks like it cost two million dollars will become normal instead of remarkable. This raises the floor for everyone, which is good. It means sloppy work will stand out more. But it also means that great storytelling on a budget won't look like a budget limitation anymore. It'll just look like great filmmaking.

What Doesn't Change

Now let me tell you what I don't think is coming, no matter how good the tools get. I don't think AI is going to direct films better than humans. I don't think AI is going to write scripts with the specific weirdness and truth of human experience. I don't think AI is going to know when to hold on a face for three seconds longer than is comfortable because that's where the truth lives.

I don't think AI will ever understand why the lead character keeps his dead wife's coffee mug on the counter, unwashed, with the stain still in it. I don't think AI knows what that object means to someone who's lost the person they love most. I don't think AI knows why that's the detail that makes audiences cry every time. That's human knowledge. That's lived experience. That's the stuff great films are built from, and no amount of computing power replaces it.

The person who wins in AI filmmaking — not next year, not in five years, but always — is the person who knows what they want and can finish the project. The tools remove barriers. They don't remove the need for taste, for judgment, for the hard work of figuring out what story you're trying to tell and why anyone should care. A great tool in the hands of someone with nothing to say still produces empty work. A mediocre tool in the hands of someone with a story that burns inside them produces something that matters.

I've seen this pattern before. Desktop publishing didn't make everyone a great graphic designer — it just made it possible for great graphic designers to work without a print shop. Digital audio workstations didn't make everyone a great musician — they just made it possible for great musicians to record without a studio. Nonlinear editing didn't make everyone a great editor. It just made editing accessible. AI filmmaking tools won't make everyone a great filmmaker. They'll make it possible for great filmmakers to emerge from places and circumstances where filmmaking was previously impossible.

That's the democratization argument, and I'm not going to dress it up in fancy language because I don't believe in fancy language. More people making films means more stories. More stories means more perspectives. More perspectives means audiences get to see experiences that aren't just the same twelve plots recycled by studios with nine-figure marketing budgets. AI filmmaking isn't about replacing Hollywood. It's about enabling the other ninety-nine point nine percent of people who have stories to tell and never had the means to tell them.

The next great film about the immigrant experience might come from someone who couldn't afford film school. The next great horror film might come from a teenager in a developing country who's never held a camera but has nightmares worth sharing. The next great love story might come from a grandmother who finally has time to write now that she's retired. That's what democratization actually looks like. Not a tool replacing a person. A tool enabling a person who would have been silent.

Pro Tip: Stay Tool-Agnostic

Don't marry your tools. I know that sounds cold, but this is a business relationship, not a romance. The workflow I've described in this book — HeyGen for characters, Seedance for environments, ElevenLabs for voice, DaVinci for editing — that's what works right now, today. Next year, some of those tools will be better and some will be eclipsed by newcomers. If you've built your entire creative identity around being a "HeyGen filmmaker" or an "ElevenLabs creator," you'll resist switching even when switching is obviously the right move.

Build systems, not tool dependencies. Your show bible works no matter what character generator you use. Your script structure works no matter what video generator you use. Your editing judgment works in DaVinci or Premiere or Final Cut or whatever comes next. The craft is portable. The tools are temporary. I've been making films long enough to see tools come and go. Avid was the only professional option once. Then Final Cut Pro changed everything. Then Premiere. Then DaVinci ate everyone's lunch by being free. The filmmakers who survived every transition were the ones who focused on craft while the tools changed around them. Be one of those.

Common Mistake: Waiting for Perfect

Every week, sometimes every day, someone in an AI filmmaking forum posts some variation of this: "I'm going to start my film when HeyGen gets better hand control" or "I'm waiting for Runway to release their new model" or "Should be a few months and then the tools will be ready."

No. Start now. Start with the limitations. Start with the hands that aren't quite right and the backgrounds that drift slightly and the voices that need a little massaging in post. Because by the time that feature ships, there will be another feature you're waiting for. And another. The perfect tool never arrives. It approaches asymptotically, always getting closer, never touching down. The filmmakers who succeed aren't the ones who waited for perfect conditions. They're the ones who made films in imperfect conditions and learned from the struggle.

the sample short was made with 2024 tools. It has imperfections I can spot immediately. the lead character's hand in the recording booth scene does something weird for two frames. The apartment window in the wide shot doesn't match the close-up exactly. Things I would fix if I were making it today. But it exists. It's screened at festivals. It led to client work that paid ten times its budget. Real people watched it and felt something. If I'd waited for better tools, I'd still be waiting and nobody would have seen it. The imperfect film that exists beats the perfect film that doesn't. Write that on a piece of tape and stick it to your monitor where you can see it every day.

The Possibilities Nobody's Talking About

What if the concept of "low budget" just disappears? Not gets smaller — disappears entirely. What if, in five years, a teenager with a story and a laptop can produce something visually indistinguishable from a forty-million-dollar studio film? I'm not saying that's definitely coming. I'm saying the trajectory points in that direction and nobody has a good argument for why it would stop. The gap between amateur and professional production value is closing faster than any industry analyst predicted. At some point — and I think it's sooner than most people think — the gap becomes irrelevant. Audiences won't know and won't care what a film cost to make. They'll care if it's good.

Budget will be a behind-the-scenes detail, like what brand of camera was used. Nobody watching a great film asks what it cost. They ask what it made them feel.

What if every high school in America has a film program that produces festival-quality work? Not a few wealthy schools with equipment donated by famous alumni. Every school. Every student with a story and the will to tell it. The creative output of a generation, finally unleashed because the barrier to entry fell to zero. We have no idea what stories we're missing right now because the people who would tell them never got the chance. There's a brilliant screenwriter working at a grocery store in Kansas who doesn't know she's brilliant because she's never had a camera. There's a natural director driving a truck in Ohio who sees visual compositions everywhere he goes but has no way to capture them. AI filmmaking is for them.

It's for everyone who was excluded by the old economics.

What if the next Spielberg is a sixteen-year-old in Mumbai who has never held a camera? What if she's already making short films on her phone, learning the tools, developing her voice, and in two years she releases something that makes the industry stop and stare? That's not wishful thinking. That's probability. Millions more people can make films now. Millions more people are making films now. Probability says some of them are extraordinary. Probability says we just haven't seen their best work yet because they haven't had enough time to develop it.

The technology improves. That's the one certainty in all of this. Everything else — what we make with it, who gets to participate, what stories finally get told, which voices we get to hear — that's up to us. That's up to you. The tools are waiting. They've never been better than they are right now. They'll be better tomorrow. But they're good enough today. They were good enough yesterday. The only variable, the only thing standing between you and the film you want to make, is what you do with them.

Chapter 12: The 90-Day AI Film Plan

You have a film. I showed you how to make it. Now I'm going to show you what to do with it. No stories this chapter. Just action. If you want stories, go back to Chapter 1. If you want results, keep reading.

I don't believe in vague advice. "Follow your dreams" is useless. "Network more" is useless. "Build your brand" might be the most useless string of words ever assembled. What I'm going to give you is a week-by-week plan with specific tasks you can check off. Do the tasks. Check them off. Move to the next week. That's how films get distributed. That's how audiences get built. That's how careers happen — one checked-off task at a time.

Print this chapter. Put it on your desk. When you finish a task, cross it out with a pen. There's something about physically crossing out a task that the digital world hasn't replicated. Trust me on this. I'm sixty-five. I know what I'm talking about.

Week One: Lock Your Foundation

Your film is done. Don't touch it anymore. This week you're building the infrastructure that will support everything that comes after.

Lock your trailer. If you haven't made one yet, cut a sixty-to-ninety-second trailer from your finished film. Not a scene. A trailer. It needs to stand alone as its own piece of art. Someone who knows nothing about your film should watch this trailer and want to see more. Test it on three people who haven't seen the film. If they don't ask "When can I watch this?" within ten seconds of it ending, recut it. A bad trailer kills more films than bad reviews.

Build your press kit. This is a PDF — yes, a PDF, professionals still use them — with your film's logline, synopsis, director's statement, cast and crew (which is you, listed as multiple roles if that's accurate, don't pretend you had a crew you didn't have), production notes, and five to eight high-quality stills from the film. If your stills look AI-generated in a bad way, generate them again until they look good. The press kit is what festival programmers, bloggers, and distributors will look at before they watch your film. Make it professional. Make it honest.

Create your filmmaker website. Not a Wix template you spent twenty minutes on. A real website with your film's page, an about page, a contact page, and a way for people to watch or request a screener. Your domain should be your name or your production company name. If you don't have a production company name yet, use your name. jimbockfilms.com or whatever your equivalent is. Buy the domain. Set up the site. It should be live by Sunday.

Grumpy Old Man Tool Note: Wix and Your Own Website

Wix and other website builders are not glamorous, but a filmmaker needs a home base. Social platforms change rules, bury posts, ban accounts, and generally behave like landlords with mood swings. Your website is the place where the film lives without begging an algorithm for attention.

Use it for your trailer, poster, synopsis, press kit, contact form, email signup, screening links, behind-the-scenes posts, and proof that you are a real person making real work. Pair it with ConvertKit, Mailchimp, or Substack and you begin building an audience you can actually reach. That matters because attention rented from platforms is not the same as attention you own.

The limitation is ugliness and neglect. A bad website looks worse than no website. Keep it clean. One film page. One about page. One contact page. One signup. Do not build a digital flea market. People should understand what you made and what to do next before they lose patience and go watch dog videos.

Set up or clean up your social media. You need at least one platform where you post regularly about your work and your process. Not your personal account where your friends see your breakfast. A professional account dedicated to your filmmaking. Instagram and X are the minimum. TikTok if you're willing to make short-form video about your process. Post your first piece of behind-the-scenes content this week. Something real. A screen recording of your generation process. A photo of your workspace. A sentence about what you learned making this film. Just start.

Register your copyright. I covered this in an earlier chapter but I'm repeating it because people skip this and regret it later. Go to copyright.gov. Register your film. It costs forty-five dollars and takes twenty minutes. Do it this week.

Grumpy Old Man Tool Note: copyright.gov

copyright.gov is not flashy, but it is one of the grown-up stops in the process. If you create a script, film, artwork, or other protectable work, registration can matter. This is the part where everybody wants to skip ahead to applause and forget that ownership is part of the business. Do not be that person.

The possibility is defense and professionalism. Registering work, keeping creation records, saving prompts, tracking licenses, storing consent forms, and documenting rights gives you a paper trail. That paper trail matters when you pitch, distribute, sell, license, or defend a project. Pair copyright records with your production bible, cloud backups, contracts, and release forms. The creative future still needs folders with dates on them.

The limitation is misunderstanding. Copyright registration does not magically solve every legal issue, especially with AI-generated materials, likeness rights, music rights, platform terms, and contracts. When the stakes are high, talk to a lawyer. I know, nobody likes that sentence. But nobody likes dental work either, and teeth still matter.

Week Two: Festival Preparation

Research ten to fifteen target festivals. Not the top five everybody submits to — Sundance, Cannes, Tribeca. Submit to those too if you have the budget and the timeline lines up, but you need a mix. Regional festivals that love local filmmakers. Genre festivals if your film fits a genre. Niche festivals focused on themes your film explores — grief, family, technology. Student festivals if you qualify. Online festivals if your film is short enough to play well on screens.

Write your director's statement. This should be two hundred to three hundred words about why you made this film and what it means to you. Not a synopsis. Not a technical breakdown. The emotional reason. Why this story? Why now? Why you? Festival programmers read these. They want to know there's a human being behind the film. Give them one.

Submit to your top three festivals — the ones with deadlines coming up soonest. Don't wait for the late deadline if you can afford the regular one. Early submissions get more attention because programmers aren't exhausted yet. Create your FilmFreeway profile if you haven't already. Upload your film, your press kit, your trailer, your stills. Make your profile complete. An incomplete profile looks unprofessional.

Grumpy Old Man Tool Note: FilmFreeway and Withoutabox

FilmFreeway is the practical doorway into most festival submissions. Withoutabox was part of the older festival-submission world, but FilmFreeway is the name most indie filmmakers deal with now. The point is simple: festivals run on systems, deadlines, fees, categories, rules, and formats. Ignore that and your film does not get rejected for art. It gets rejected because you failed paperwork kindergarten.

The possibility is strategic targeting. Submit the right film to the right festivals with the right materials: logline, synopsis, director statement, poster, trailer, stills, runtime, and contact information. Use your AI tools to help draft better copy, build a cleaner press kit, and create alternate submission language. This is not just administration. This is how the film introduces itself before anyone presses play.

The limitation is cost and fantasy. Do not submit to fifty festivals because you saw a laurel on someone else's poster and got excited. Pick smart. Track deadlines. Watch fees. Read rules. A festival strategy is not buying lottery tickets while wearing a beret. It is a business decision with popcorn nearby.

Week Three: Distribution Setup

Upload your film to YouTube and Vimeo. Use optimized titles — not "My Film Final Version 3" but the actual title of your film. Write descriptions that include relevant keywords without being spammy. Tag appropriately. Set the thumbnail to one of your best stills. The thumbnail matters more than you think. It's the poster for your film on these platforms.

Grumpy Old Man Tool Note: YouTube and Vimeo

YouTube and Vimeo are where a finished film can finally leave the hard drive, which is important because a film nobody can watch is not a film. It is a secret with a render file. YouTube gives reach, search, comments, and discoverability. Vimeo gives a cleaner professional presentation and better control for screeners, clients, and private links.

Use both intelligently. YouTube for public audience, trailers, behind-the-scenes, process videos, shorts, and search traffic. Vimeo for festival screeners, client review links, pitch pages, and clean embeds on your own website. Pair them with your email list so viewers are not just random numbers on a dashboard. Numbers are nice. Names and emails are better.

The limitation is platform dependency. YouTube can bury you. Vimeo can sit there looking elegant while nobody visits. Uploading is not marketing. It is just putting the baby on the porch. You still have to tell people the baby exists.

Embed the YouTube or Vimeo player on your website's film page. Test it. Make sure it works on mobile. Half your audience will watch on their phones. If the embed is broken or the player is tiny on mobile, fix it. Test it again.

Create a behind-the-scenes page on your website. Call it "How I Made This" or "The Process" or whatever fits your voice. This page should document your workflow — what tools you used, what you learned, what went wrong, what surprised you. Include screenshots. Include failed generations alongside successful ones. This page serves two purposes: it builds credibility with audiences who are curious about AI filmmaking, and it demonstrates expertise to potential clients who might hire you.

Set up an email list. I use ConvertKit. Other people use Mailchimp or Substack. Doesn't matter which. What matters is that you have a way to contact people who want to hear from you. Offer your film as a free screener link in exchange for an email signup. Not the YouTube link — a private Vimeo link or a direct download. Something that feels exclusive. Put the signup form on your website. Promote it on your social media. Get your first ten subscribers this week even if you have to ask friends and family individually.

Grumpy Old Man Tool Note: ConvertKit, Mailchimp, and Substack

ConvertKit, Mailchimp, and Substack are audience tools, and audience tools matter because social media is a casino with better lighting. An email list is boring in the best possible way. You collect permission to contact people who care about your work, your process, or your next film. That is worth more than a thousand drive-by likes from strangers who forgot you existed four seconds later.

Use these tools to build a filmmaker newsletter: production updates, trailer drops, festival announcements, behind-the-scenes notes, lessons learned, early access, and release reminders. Combine the list with your website, YouTube/Vimeo links, and social posts. The future filmmaker is not just a maker. The future filmmaker is a small media company, whether he likes it or not.

The limitation is neglect. Do not start a newsletter and then send one message every eight months that begins with 'Sorry I have not written.' Nobody cares. Pick a pace you can keep. Monthly is fine. Useful is better than frequent. And for the love of all that is decent, do not turn every email into a desperate sales flyer.

Week Four: Audience Building Begins

Post behind-the-scenes content every day this week. I mean every day. One piece of content minimum. A screenshot with a caption about your process. A before-and-after of a generated shot. A voice recording of you talking about why you became a filmmaker. A clip from your film with context about how you made it. Daily posting sounds like a lot because it is. It takes thirty minutes a day. You have thirty minutes.

Engage with other filmmakers' work sincerely. Watch their films. Leave thoughtful comments. Share work you admire. The AI filmmaking community is still small enough that genuine engagement gets noticed. The people you support this week will support you next week. This isn't networking in the gross transactional sense. It's being a member of a community. Do it genuinely or don't do it at all.

Join AI filmmaking communities if you haven't already. Discord servers, Facebook groups, Reddit communities, whatever platform you prefer. Introduce yourself. Share your work when appropriate. Answer questions from people who are newer than you. Teaching reinforces what you've learned. Plus it's the right thing to do.

Grumpy Old Man Tool Note: Discord, Facebook Groups, and Reddit

Discord, Facebook Groups, and Reddit are where communities gather, argue, compare tools, share clips, ask questions, and occasionally act like raccoons fighting over a sandwich. Used wisely, they are valuable. Used foolishly, they are a time sink with notifications.

The possibility is learning and audience development. Join AI filmmaking groups, ask real questions, share process, post before-and-after tests, document mistakes, and become useful. Useful people get remembered. Show-offs get muted. These communities can lead to collaborators, early viewers, festival tips, tool discoveries, and client leads if you behave like a human being instead of a walking advertisement.

The limitation is noise. Every community has people who speak with great confidence after using a tool for seventeen minutes. Listen, but verify. Try the workflow yourself. Do not build your film around forum rumors. The internet is full of unpaid experts with imaginary trophies.

Reach out to three film bloggers or podcasters who cover indie film or technology in film. Not the biggest ones — the mid-size ones who actually respond to emails. Write a personal message. Mention something specific they've written or said. Briefly introduce your film and ask if they'd be interested in a screener. Don't follow up more than once if they don't respond. Move on. It's not personal. They're busy.

Week Five: Business Development

Create a services menu. You're a filmmaker now. That means you can offer filmmaking services. Corporate videos, promotional content, real estate tours, training videos, nonprofit storytelling — businesses need video and most of them have no idea AI filmmaking exists. List your services with clear descriptions and clear pricing. Don't make people guess. "Short promotional film: starting at $1,500." "Corporate training video: starting at $2,000." You can adjust these numbers based on your market. But have them written down.

Reach out to ten potential clients. Local businesses, nonprofits, startups, anyone who might need video content. Email them personally. Reference their specific business. Don't send a form letter. Mention that you use AI-assisted filmmaking techniques that allow you to deliver professional results at a fraction of traditional costs. That's your competitive advantage. Use it.

Follow up on any festival acceptances or rejections. If you got accepted, promote it everywhere. If you got rejected — and you will get rejected, everyone does — submit to the next festival on your list. Rejection is information, not verdict. It tells you one festival on one day didn't select your film. That's all it tells you. Move on.

Attend one local industry event this week. A film screening, a networking meetup, a workshop, a panel discussion. Show up in person. Introduce yourself to three people you don't know. Have a business card or a phone with your website ready to show. The film industry still runs on personal connections. Online communities are great. In-person connections are better.

Week Six: Double Down on Audience

Post your "How I Made This" content. A video essay, a blog post, a Twitter thread, a carousel on Instagram — whatever format fits your platform. Make it detailed. Make it honest. Show the failures alongside the successes. People are fascinated by process, especially AI filmmaking process because it's so new. Your documentation is content.

Grumpy Old Man Tool Note: TikTok, Instagram, Facebook, X, and LinkedIn

TikTok, Instagram, Facebook, X, and LinkedIn are not the same animal. TikTok rewards fast hooks and process clips. Instagram likes visual polish and reels. Facebook still reaches certain groups and older audiences. X can spread a strong thread if the wind is blowing right. LinkedIn is where business people pretend not to be on social media while looking for vendors. Use each platform for what it actually is, not what you wish it was.

The possibility is turning production into marketing. Show the AI setup. Show the before and after. Show the prompt, the bad generation, the fix, the final shot. Show how one scene went from idea to voice to image to edit. The audience does not only want the finished film. They want to watch the magician build the trick, especially when the trick costs less than a used lawn mower.

The limitation is addiction. Social media will gladly steal the time you needed to finish the film. Post with purpose. Repurpose content. Batch clips. Link back to your website and email list. Do not spend three hours arguing with a stranger named FilmTruth472 about whether AI is real cinema. That man is not buying your book and he is not funding your movie.

Reach out to three more bloggers or podcasters. The first three might not have responded. That's fine. Try three more. Adjust your pitch based on what you've learned. Keep refining. The people who persist are the people who get featured.

Start a newsletter if you haven't already. Weekly or biweekly. It doesn't have to be long. Five hundred words about what you're working on, what you've learned, what you've watched that inspired you. The newsletter builds a direct relationship with your audience that doesn't depend on social media algorithms. Someone who gives you their email is someone who wants to hear from you. Respect that. Show up consistently.

Review your analytics. What's working on social media? What's driving traffic to your website? What's getting ignored? Do more of what works. Do less of what doesn't. This isn't complicated. Most people don't review their analytics because they'd rather post and hope. Don't be most people.

Week Seven: Reflect and Plan

Take a breath. You made a film. You distributed it. You built infrastructure. You started building an audience. That's more than most people who call themselves filmmakers ever do. Acknowledge that. My wife would tell me to stop here and appreciate what I've accomplished before rushing to the next thing. She's usually right about this stuff.

Reflect on what you learned from Project One. Write it down. What worked? What didn't? What will you do differently next time? What tools do you want to explore? What stories do you still want to tell? This reflection document becomes the foundation for Project Two. Don't skip it. The learning compounds only if you capture it.

Research what other AI filmmakers are doing. Watch their films. Note what's working in their distribution strategies. Join conversations. Ask questions. The community is moving fast and there's something new to learn every day. Spend this week being a student again.

Week Eight: Project Two Development Begins

Write the logline for your next film. One sentence. If you can't say it in one sentence, you don't know what it is yet. Keep refining until you have a sentence that makes people say "I'd watch that." Test it on three people. Their reaction is your data.

Create a mood board. Visual references, color palettes, shot compositions, music that captures the feeling, other films that share your intended tone. This isn't homework. This is the fun part. Collect images that make you feel the way you want your audience to feel. Put them somewhere you can see them every day.

Start training your new Digital Twins. If you're continuing a series, you might reuse existing characters with new outfits. If you're starting fresh, begin the training process for your new cast. This takes time. Start it now so you're not waiting on it later.

Lock your script outline. Not the full script — the outline. Beginning, middle, end. Three-act structure or whatever structure fits your story. Know where you're going before you start writing scenes. The outline is your map. The script is the journey. Don't start the journey without a map. I've tried it. You get lost.

Week Nine: Keep Building

Continue posting content daily or at least five times a week. Your momentum matters. An audience that hears from you regularly will pay attention when you announce your next project. An audience that hears from you sporadically will forget you exist.

Follow up with the clients you reached out to in week five. Some will have responded. Some won't. Follow up once, politely, referencing your original message. Then move on. Don't chase. There are always more potential clients. Your time is better spent making your next film than chasing someone who isn't interested.

Submit to three more festivals. Keep the pipeline full. Festival programming cycles are long. A rejection this month doesn't preclude an acceptance next month. Keep submitting until you've exhausted your target list or your budget.

Week Ten: Community and Collaboration

Help another filmmaker. Seriously. Find someone who's newer than you and answer their questions. Review their work. Give honest, kind feedback. The community you build by helping others is the community that sustains you when you need help. This isn't altruism — though it is that too. It's strategy. The filmmakers you help today will remember you tomorrow.

Document your process. Start collecting screenshots, notes, and reflections for a case study of your first film's distribution. This becomes content you can share, teaching material you can use, and a record you can learn from. Documentation is a habit. Build it now.

Reach out to one potential collaborator. An actor who might want to be the voice model for a character. A musician who might score your next film. A writer who might co-write with you. You can make films alone — I made the sample short essentially alone — but collaboration makes the work richer. Find one person whose skills complement yours. Start a conversation.

Week Eleven: Refine Your Systems

Review everything you've built in the past ten weeks. Your website. Your social media presence. Your email list. Your festival submissions. Your client outreach. What's working? What's broken? What's half-finished? Fix the broken things. Complete the half-finished things. Let go of the things that aren't working.

Write the first draft of your Project Two script. Get it on paper. It will be bad. First drafts are always bad. The goal isn't a good script — the goal is a complete script you can revise. You can't revise a blank page.

Set three goals for the next ninety days. Not twelve goals. Three. Specific and measurable. "Finish Project Two script" is a goal. "Get better at filmmaking" is not. Write them down. Put them where you'll see them.

Week Twelve: Lock and Load

Finalize your pre-production for Project Two. Lock the script — not perfect, but locked. Confirm your Digital Twins are trained and consistent. Prepare your show bible. Set your production schedule. Know what you're generating first, second, third.

Celebrate what you've accomplished. In twelve weeks you made a film, distributed it, built infrastructure, started an audience, and launched your second project. That's not nothing. That's more than most. My wife would insist I take her out to dinner at this point, and she'd be right. You need to mark the milestones or the work just becomes an endless treadmill.

Then start generating scenes for Project Two. Because the work continues. And the work is the point. Making films is what filmmakers do. Everything else — festivals, audiences, clients, recognition — that's a byproduct of making films. Keep making them.

Pro Tip: Calendar Everything

Set calendar reminders for every weekly milestone. Not a to-do list — actual calendar entries at specific times. "Tuesday 9 AM: Post behind-the-scenes content." "Thursday 2 PM: Follow up with client outreach." "Sunday evening: Review analytics." When a task has a specific time slot, it gets done. When it's on a list without a time, it gets procrastinated indefinitely. I've learned this the hard way over seven decades of doing things and not doing things. The calendar is your friend. Use it religiously.

Find an accountability partner. Another filmmaker, not your spouse. Someone who understands the work and will call you out when you're slacking. Check in weekly. Share what you accomplished and what you didn't. The social commitment of reporting your progress to someone else is weirdly powerful. Use it.

Batch similar tasks. Do all your festival submissions in one sitting. Write all your social media captions for the week in one block. Generate all your voice work in one session. Context switching kills productivity. Batching preserves it. I do all my HeyGen generation on Mondays, all my voice work on Tuesdays, all my editing Wednesday through Friday. Saturday I handle business. Sunday I rest or my wife makes me rest. That rhythm works for me. Find your rhythm.

Common Mistake: The Week One Sprint

I've seen this pattern a hundred times. Someone reads a plan like this and tries to do all twelve weeks in the first week. They submit to fifteen festivals, build a website, start a newsletter, reach out to twenty clients, and burn out by day four. Then they do nothing for three months because they're exhausted.

Don't do this. The plan is paced for a reason. One week at a time. One task at a time. The tortoise wins this race. The hare gets a lot of attention in the first week and then disappears. I've been the hare. Be the tortoise.

Another common mistake: skipping the legal and business setup because it's boring. Copyright registration isn't exciting. Website building isn't exciting. Email list setup isn't exciting. But these are the foundations that everything else builds on. Skip them and you'll find yourself six months from now with a successful film and no way to capitalize on it. Do the boring work. It's what separates professionals from hobbyists.

Don't compare your progress to other filmmakers. They're on a different timeline with different resources and different starting points. The filmmaker who posts about their Sundance acceptance might have been making films for fifteen years. They might have family money. They might know someone who knows someone. Their path isn't your path. Your only meaningful comparison is to yourself last month. Are you further along than you were? That's the only metric that matters.

And please, for the love of all that is holy, don't give up because one festival rejected you. Rejection is the default state of filmmaking. Every working filmmaker has been rejected more times than they can count. The ones who succeed aren't the ones who never get rejected. They're the ones who keep submitting, keep making work, keep showing up. Be one of those.

Your 90 Days Becomes Your 900 Days

Here's what I want you to understand as we close this book. This ninety-day plan isn't just about one film. It's about building a system that produces films, builds audiences, and generates income — indefinitely. Do this plan once and you've distributed one film. Do it twice and you've built a catalog. Do it three times and you're a working filmmaker with a body of work, an audience that follows you, and probably some income from client work or distribution deals.

The film you made in these ninety days might become the calling card for everything you'll ever do. the sample short was that for me. Twelve minutes made for $847. It led directly to client work that paid ten times its budget. It got me invited to speak at events. It introduced me to collaborators I'm still working with. That one small film — the first one I finished, the one I almost gave up on three times — opened every door that came after.

Your film can do the same. Not because it's perfect. Not because it wins every festival. Because you finished it. Because you put it out into the world. Because you built the infrastructure to support it and the discipline to keep going. The filmmakers who succeed aren't necessarily the most talented. They're the ones who finish things and release them. That's it. That's the secret. There's no more to it.

I started this book by telling you that I'm sixty-five years old and I've been making films for over a decade. I'm not a tech genius. I'm not a Hollywood insider. I'm a guy from New Jersey who figured out that AI tools let him tell stories he couldn't tell any other way. I made the sample short in my kitchen. I made the second sample short for about thirty dollars. I built a career from those small beginnings, one film at a time, one checked-off task at a time.

You can do this. Not in some motivational poster way. In a real, practical, specific way. The tools are on your computer or they will be by tonight. The knowledge is in this book and in the communities you can join right now. The only remaining question is whether you'll do the work.

So do the work. Make the film. Distribute it. Build the audience. Make the next one. Keep going. I'll see you at the festival.

SECTION 3: RIGHTS, AUDIENCE, AND SELLING YOUR FILM

Marketing, Distribution, and Protection

The reader enters this section with a finished film and asks: how do I get people to watch it, how do I protect it, and how do I build a life around making more? By the end, they have a complete distribution strategy, legal framework, and business model for sustainable AI filmmaking.

Chapter 13: The Trailer as Seduction

My wife and I went to see a movie last fall. I won't name it because the poor bastard who made it probably has enough problems. But the trailer had us both hooked — explosions, a beautiful woman running through rain, some line about "the truth can wait." Sold. We bought tickets, got our popcorn that costs more than my first camera, settled in. Two hours later we walked out and I turned to her and said, "That movie had nothing to do with the trailer." She nodded. The trailer was a horror film. The actual movie was a slow family drama about inheritance law. Somebody in marketing had looked at a hundred hours of footage, found the one scene where a car backfired, and built an entire action trailer around it. That's fraud, basically.

But here's the thing — that trailer worked. We bought tickets. And sitting there in my anger, I realized something that changed how I approach every film I make now: the trailer is not a summary. It's not even really about the film. The trailer is a seduction. It's a promise of a feeling. And if you get that feeling right, people will show up. If you get it wrong, or worse, if you bore them with information, you might as well throw your film in the Passaic River.

This hit me hardest when I was finishing the sample short. I'd spent months on this thing — $847, every Saturday, learning the tools, obsessing over the lead character's navy sweater and the amber light on his mixing board. I exported the final version at 2 AM on a Tuesday. I remember staring at the file on my desktop: FatherFrequency_Final_v7.mp4. It was beautiful. It was done. And I had absolutely no idea how to get anyone to watch it. My first instinct was the same instinct that ruins most indie filmmakers: I made a "trailer" that was basically a chronological summary. Here's the lead character in his studio. Here's a supporting character in her kitchen. Here's the child character confronting him. Here's the emotional climax. Boom — title card. I showed it to my wife.

She watched it, smiled that polite smile she uses when I've overcooked the chicken, and said, "It tells me everything that happens." She was right. It was a Wikipedia entry with music. Nobody was going to watch that and feel anything except the satisfaction of having the plot explained to them.

So I started over. I asked myself a different question. Not "what happens in this film?" but "what does this film feel like at 3 AM when you're alone and something just hit you?" The answer was loneliness. The answer was a man surrounded by sound who couldn't hear his own life anymore. That was the trailer. Not the sister, not the daughter, not even the dead wife's voice. Just the lead character, alone, at his mixing board, at 3 AM, the amber light making him look like a ghost in his own house. I opened with that shot. Held it for five seconds. No music. Just the hum of the room. Then I cut to his hand on a fader. Then to his wedding ring catching the light. Then a flash of the absent wife's photo on the shelf — just a glimpse, barely there. That was the hook.

Not information. A feeling.

What a Trailer Actually Is

If you're coming from the edit room like I was, you've spent weeks — maybe months — making sure your story makes sense. Every beat is earned. Every transition is smooth. Every emotional moment has setup and payoff. Great. Now forget all of that. The trailer exists in a different universe. It's not the film's servant. It's the film's doorbell. Its job is to make someone curious enough to open the door.

Think of it like a diner menu. The best diners — the ones in Jersey where the coffee is always hot and the waitress calls you hon — they don't describe every ingredient. They say "meatloaf like your mother wished she made." That's the trailer. It's not the meatloaf. It's the promise of what the meatloaf will do to you. If your trailer tries to be the full meal, it fails. If your trailer makes someone hungry, it wins.

The structure that worked for me — after a lot of trial and error, and after showing drafts to Chuck the plumber who honest-to-God fell asleep during version three — is a simple three-part arc. Hook. Escalation. Emotional payoff. Sixty to ninety seconds. That's it. If you can't sell someone on your film in ninety seconds, you don't have a film, you have a home movie.

The hook is the first five to ten seconds. This is where most indie trailers die. They open with a title card. "NJ Film Productions Presents." Nobody cares. Then maybe some establishing shot of a city. Then a fade-in to dialogue. Wrong. All wrong. You know what the best hook is? The best image you've got. The shot that made you gasp when you first saw it come out of the generator. The moment that surprised even you. For the sample short, that was the lead character alone at 3 AM — not because it was the most dramatic shot, but because it asked a question. Who is this man? Why is he alone? Why do I feel sad looking at him? Questions are hooks. Answers are not.

The escalation builds tension without revealing resolution. You show relationships under pressure. You show faces reacting to things the audience can't see yet. You show the promise of conflict, not the resolution of it. In The sample short's trailer, the escalation was three quick shots: the child character's back as she stands in the studio doorway (she hasn't turned around yet — you don't know who she is), a supporting character's hand reaching for a photo on the kitchen table (you don't see whose photo), and the lead character pressing play on a recorder, his face crumbling just a fraction before the cut. Three shots. Maybe six seconds total. But they build a sense that something is coming apart.

The emotional payoff is the final fifteen to twenty seconds. This is where you give them one clean hit of feeling — a line of dialogue that lands, a musical swell, a face in close-up that breaks your heart. For the sample short, I used the absent wife's voice. Just her voice. "the lead character. You're still wearing the sweater." Cut to black. Title. That's the entire payoff. You don't know who the absent wife is. You don't know if she's alive or dead or a recording or a dream. But you feel the weight of a marriage in that one line. That's the promise. That's what makes someone click "watch the full film."

The Shots You Use (and the Shots You Don't)

Here's something that took me too long to learn: your trailer does not have to use only shots from the finished film. I know, it sounds obvious now, but I was trapped in the mindset that a trailer is "scenes from the movie." It's not. A trailer is marketing. And marketing can use anything that sells the promise.

For the sample short, I generated four additional shots specifically for the trailer. They don't appear in the film. One was an extreme close-up of the lead character's eye, the amber light reflected in the iris, a tear forming but not falling — just held there. Another was a slow push into the absent wife's photo on the shelf, the frame slightly askew. A third was the mixing board in complete darkness, every LED light burning like a tiny star. And the fourth was a supporting character, shot from behind, standing in a doorway that leads to warm light — you can't see her face, just the shape of a woman hesitating.

These shots came from Seedance, my B-roll generator. I prompted for trailer-specific imagery. "Extreme close-up of weathered man's eye, amber desk lamp reflected, emotional, cinematic, shallow depth of field." "Dark mixing board with glowing LED lights, atmospheric, lonely." Each took maybe ten minutes to generate and refine. None cost extra because I was already subscribed. And they transformed the trailer from "clips from the movie" into "a visual poem about grief and sound."

This is where AI filmmaking gives you a massive advantage over traditional filmmakers. A studio shooting on location can't easily go back and get "one more shot of the actor's eye with different lighting." They'd need the actor, the crew, the location, insurance, catering, three weeks of scheduling. I needed a text prompt and a coffee refill. I generated six versions of that eye close-up with slightly different prompts until I got the one that made my stomach flip. That's the shot that opens the trailer now. the lead character's eye, the amber light, a tear held like a held breath. It tells you everything about the film before you hear a single line of dialogue.

the second sample short was the opposite problem. That film is four minutes long. Cost about thirty bucks. It's basically one long joke with three punchlines. If the trailer showed any of the pranks — the whoopee cushion, the dirty limerick, the coffin recording — the movie would be ruined. So the trailer showed zero pranks. None. Not a hint. Instead, it was just Harold's face. Sixty seconds of watching a mid-sixties funeral director in a black suit get progressively more disheveled while classical organ music played underneath. First shot: Harold in a pristine black suit, tie straight, every hair in place. He looks at the camera with exhausted dignity.

Second shot: tie loosened, one sleeve rolled up, he's staring at something off-screen like his life is collapsing. Third shot: jacket abandoned, both sleeves rolled, he's sitting on church steps holding a whoopee cushion with an expression of pure existential dread. Final shot: undershirt, tie completely gone, dirt on his cheek, and the faintest smile. Cue title. Cue record scratch.

That trailer told you everything and nothing. You knew it was a comedy about a funeral. You knew the main character was going through something. You knew the tone was absurd and dark. But you had no idea about the pranks. People who watched the trailer told me later they expected a melancholy character study. Then the whoopee cushion hit and they laughed so hard they choked on their popcorn. The trailer set them up. The film knocked them down. That's the game.

Music: The Secret Weapon

The music you choose for your trailer determines how people will feel about your entire film. This is not hyperbole. I tested The sample short's trailer with three different music tracks. Same visuals. Same edit. Track one was the actual score from the film — a quiet, melancholy piano piece I'd generated with Suno. It was beautiful. It was also wrong for the trailer. The trailer felt slow. Contemplative. Like homework. Track two was an epic orchestral piece from a royalty-free library. Crescendos. Timpani. The full Hollywood treatment. Suddenly my twelve-minute film about a grieving sound engineer felt like a Marvel movie. Ridiculous. Track three was a simple ambient drone with a single piano note that repeated every four seconds, like a heartbeat.

Underneath it, barely audible, the hiss of analog tape. That was it. No melody. No crescendo. Just pulse and hiss.

That was the one. It made every visual feel heavier. the lead character's eye close-up went from "nice shot" to "I can't breathe, this man is in pain." The music didn't tell you what to feel — it created a space for you to feel something. That's what trailer music should do.

I generated that track with Udio in about fifteen minutes. "Ambient drone, analog tape hiss, single piano note repeating, emotional, empty room feeling, no percussion, no melody." The result needed two iterations to get the hiss level right — first version had too much, sounded like a broken air conditioner. Second version was perfect. I exported it, dropped it into DaVinci Resolve, and watched the trailer come alive. The whole thing — trailer edit, custom shots, music — took about four hours. Four hours to create the single most important marketing asset for a film I'd spent months on. Seems disproportionate, but that's the job. You don't spend four hours and then stop.

You spend four hours, show it to everyone you trust, listen to what they say, and spend four more. I made seven versions of that trailer before I was happy.

Pro Tip: Generate trailer footage during production, not after. Every time you're working on a scene, pause and think: what shots here could serve the trailer? A character close-up with an intense expression. A reaction shot you didn't use in the final cut. A detail of a hand or an object that reads as pure emotion without context. I now keep a "trailer bin" in DaVinci Resolve from day one of editing. Anything that gives me a feeling gets dropped in there. By the time I'm ready to cut the trailer, I've got forty or fifty shots to choose from instead of trying to pull from the finished film like a desperate man raiding his own fridge at midnight.

Pro Tip: Draft your trailer before the film is finished. I know this sounds insane. You don't even know the final pacing yet. But here's the magic: when you try to sell a story in sixty seconds, you find out real fast where your pacing problems are. I cut a rough trailer for the sample short after I'd assembled the first rough cut — before color grading, before the final music was done, before anything was polished. And I immediately saw that the middle section dragged. Two scenes that I thought were essential turned out to be redundant when compressed into trailer form. I went back and cut one of them entirely. The film got better because I tried to sell it before it was done. The trailer became a diagnostic tool.

What Most People Get Wrong

The most common mistake I see in indie film trailers is showing too much plot. Filmmakers are terrified that if they don't explain the story, people won't understand. But understanding is not the goal. Feeling is the goal. Nobody watches a trailer and thinks "I need a detailed summary of the narrative structure." They watch a trailer and think either "I want to feel that" or they don't think at all and they click away. Every plot detail you include is a bullet in your own foot. The

the sample short trailer never explains that the absent wife is dead. Never explains that the child character is the lead character's daughter. Never explains that the audio recordings are the absent wife's final messages. It just shows you a lonely man, some fractured relationships, and a voice that makes him cry.

That's enough. That's more than enough. That's everything.

Another common mistake is using the same music in the trailer that you use in the film. This feels logical — consistency, right? But the trailer and the film have different jobs. The trailer needs to compress emotion into seconds. The film has minutes — sometimes hours — to unfold. Your film's score is designed for breathing room. Your trailer music needs to be concentrated. Like espresso versus a cup of coffee. Both are coffee. One will wake you up in three sips.

Making the trailer longer than two minutes is filmmaking suicide. I don't care if your film is a masterpiece that redefines cinema. If your trailer is two and a half minutes, you've already lost. People's attention spans are not a resource you can negotiate with. They're a wall you hit. Sixty to ninety seconds. That's your window. If you can't make someone feel something in ninety seconds, you need to go back and look at your film harder. Something is missing.

And for the love of all that is holy, do not open with a title card. I will drive to your house and unplug your computer. Open with your best image. The thing that made you fall in love with your own film. The shot that surprised you. The moment that felt like magic. That's your opening. Not your production company's logo. Not "a film by." Not a black screen that fades slowly into something. Hit them. Hard. In the first three seconds. Every frame before your best image is a frame you're asking someone to sit through on faith. And faith is in short supply on the internet.

Common Mistake: Using dialogue exposition in the trailer. I see this constantly. Some character says "We need to talk about what happened" and then the trailer cuts to a reaction shot and then someone else says "I can't believe he's really gone" and then a shot of a funeral. Stop it. Every line of dialogue in your trailer should be able to stand alone as a poem. "You're still wearing the sweater." That's the only line of dialogue in the the sample short trailer. It means nothing out of context. In context, it means everything. That's the bar. One line. Make it count.

Common Mistake: Making the trailer look better than the film. I know, I just told you to generate custom shots and use concentrated music and craft something beautiful. But if your trailer is a cinematic masterpiece and your film looks like it was shot on a potato by comparison, you've created a trust problem. The trailer should be an accurate promise of the film's quality and tone. The sample short's trailer looks like the sample short because the same tools made both. The same HeyGen avatars. The same Seedance environments. The same aesthetic world. The trailer is a door that opens into a room that matches. That's the goal.

What If You Went Wild With This?

Here's where I start getting excited and my wife tells me to lower my voice because I'm in the kitchen and she's trying to read.

What if your trailer was a completely different genre than your film? I'm serious. What if The sample short's trailer was cut like a horror movie? Quick cuts. Distorted audio. the lead character's face flashing in strobe light. the absent wife's voice pitched down to a demonic growl. "YOU'RE STILL WEARING THE SWEATER." You'd get a completely different audience. Horror fans would show up to a grief drama. Some would be angry. But some — the ones who actually watch it — would have an experience they never planned on having. And they'd talk about it forever. The trailer as bait-and-switch, but the switch is good. The trailer as Trojan horse. I haven't done this yet, but I'm itching to.

There's a version of the second sample short trailer cut as a somber meditation on mortality that makes people cry — and then the whoopee cushion happens in the actual film and their emotional whiplash becomes the story they tell their friends.

What if you made five different trailers for five different audiences? One trailer for drama fans — emphasize the grief, the marriage, the emotional weight. One for tech people — emphasize the sound engineering, the recording equipment, the craft. One for AI filmmakers — emphasize that this was made with AI tools and show the process shots as part of the marketing. One that's just music and images, no dialogue at all, pure visual poetry. And one that's sixty seconds of behind-the-scenes footage showing how you made it, warts and all, ending with the final shot. Each trailer goes to a different platform, targets a different community, speaks a different language. The film is the same. The promise is different.

That's five times the reach for the same film. And with AI tools, generating five variations is not five times the work. It's maybe twice the work. The shots are already made. You're just recombining them. You're remixing your own film like a DJ.

And here's the one that keeps me up at night: what if the trailer became its own short film? A trailer that's not trying to sell a longer piece but exists as a complete artistic statement on its own. Sixty seconds that don't point to anything beyond themselves. I've been experimenting with this — making trailers for films I haven't made yet, films I might never make. Just the trailer. Just the promise. There's something pure about it. A distillation. The haiku version of cinema. I showed one to Chuck the plumber — a trailer for a film about a man who builds a boat in his basement that will never fit through the door — and he didn't say anything for thirty seconds.

Then he said, "That made me sadder than most movies I've paid for." It was forty-seven seconds long. That's the power. When you strip away the obligation to explain, to resolve, to satisfy — when you just make the promise and let it hang there — you tap into something almost primal. The audience becomes a participant in the unfinished story. They fill in the gaps. They become the filmmaker. And they remember you because you trusted them enough to leave them wanting more.

The trailer is the first creative act of your film's life in the world. Treat it like that. Not like homework. Not like a formality. Like art. Like a seduction. Like the reason someone will choose to spend twelve minutes — or four minutes, or ninety minutes — with something you made instead of scrolling past it like they scroll past everything else. The trailer is your hello. Make it worth hearing.

Chapter 14: The Audience You Build Before the Film Is Finished

Frank from next door had a barbecue last July. He spent three hundred dollars on ribs, bought two new folding tables from Home Depot, hung string lights in his maple tree. Looked beautiful. I walked over at six with a bottle of wine, expecting a crowd. There were four people. Frank, his wife, me, and a guy who I think was Frank's coworker but might have been just some dude walking by who smelled meat. Frank stood at that grill like a captain going down with his ship, flipping ribs for an audience that wasn't there. Around eight he came over, sat in the folding chair next to me, and said, "I don't get it. I made the food. I set everything up. Where is everybody?" I looked at him.

"Frank," I said, "when did you tell people about this?" He thought about it. "This morning. I put a note on my door." I didn't say anything else. What was there to say? Frank thought making something good was enough. He thought the ribs would speak for themselves. The ribs were delicious, by the way. But nobody knew about them. And if nobody knows, nobody comes. That's not just barbecues. That's everything. That's films.

I made four films before the sample short that almost nobody watched. I'm not talking about ten people instead of a hundred. I'm talking about me, my wife, and a guy in a Discord server who clicked the link by accident and stayed because he thought it was something else. I would finish a film — pour everything into it — export it, upload it, post once on Facebook with a link, and then wait. I might as well have been Frank with his note on the door. The film was good.

The ribs were delicious. But I had built zero audience before the film existed. I thought that was how it worked: you make something great, and word gets around. Like in the movies. Like some magical fairy dust of quality that makes crowds form spontaneously. That is not how it works.

That has never been how it works. And if you think AI filmmaking changes this — if you think the novelty of AI-generated cinema means people will flock to you without effort — I have some ribs you should try.

The Audience Comes First

The hardest lesson for any filmmaker, and especially for AI filmmakers who can produce work faster than traditional creators, is this: audience building is not something you do after the film is done. It's something you do before. During. Alongside. The audience and the film grow together, and if you try to separate them — film first, audience second — you end up like Frank. Great ribs. Empty yard.

When I started the sample short, I had about a hundred followers across all platforms. Mostly people who knew me from other failed projects or from local film meetups where I'd shown up once and talked too long about digital avatars. A hundred people. That's a barbecue with folding chairs that don't even get used. But I made a decision before I generated a single frame: I was going to document the entire process publicly. Every step. Every failure. Every weird output from HeyGen that looked like a horror movie by accident. I was going to let people watch me make the film, and in doing so, I was going to give them a reason to care about it before it existed.

The first post I made was just a screen recording of me writing the prompt for the lead character's avatar in HeyGen. No music, no editing, just raw footage with me narrating. "Okay, so I need a man in his late forties who looks like he hasn't slept well in two years. Salt-and-pepper stubble. Crow's feet. Navy sweater with these leather elbow patches that my uncle used to wear. Let's see what happens." I hit generate. The first result looked like a catalog model for men's business casual. Too polished. Too alive. I laughed on camera — genuine laughter, because it really was funny — and said, "That's not the lead character. That's the lead character's successful twin who went to business school. Let's try again." Second result was closer. Third result was him.

the lead character, staring back at me from the screen with eyes that had seen something. I saved that recording and posted it. Two minutes of raw process. No finished product. Just a guy in New Jersey trying to make a fake person look sad enough.

That post got three thousand views in the first day. Not because the film was good — there was no film yet. Because the process was interesting. Because watching someone use AI to create a character with a specific emotional register felt like watching a new kind of magic trick. People commented with their own prompts. They suggested details — "give him a wedding ring that's slightly loose" and "make the stubble patchier on the left side like he only shaves when he remembers." I incorporated their suggestions. I posted the updated the lead character. They saw their input change the character. Now they were invested. Not in a finished film — in the making of it.

What to Share and Where

By the time the sample short was finished, that single account had grown to just over two thousand followers. Two thousand people who had watched the lead character come to life, who had seen a supporting character's kitchen get generated in Seedance, who had heard the first rough version of the absent wife's voice and told me it sounded too young, who watched me iterate the music through four versions before landing on the ambient drone. They weren't just an audience waiting for a film. They were collaborators. Witnesses. People with a rooting interest in something they'd helped shape.

This is the thing that traditional film schools don't teach you because traditional film schools are built around a model where you work in secret for years and then premiere at a festival and hope for distribution. That model is broken for indie filmmakers and it's especially broken for AI filmmakers, who have the unique advantage of being able to produce and share iteratively. You can show your work every day. You can make your process the content.

The platform strategy that worked for me — and I'm not saying this is the only way, but it's what I did and it worked — was to think about each platform as a different room in a house. YouTube is the living room. That's where the long-form behind-the-scenes content lives. I made twenty videos during the production of the sample short — each one ten to fifteen minutes, showing a different aspect of the process. The avatar generation episode. The voice acting episode where I tested six different ElevenLabs voices for the absent wife. The environment design episode where I built the lead character's studio in Seedance and kept getting results that looked like a spaceship until I refined the prompts. These videos weren't polished.

They were me, at my desk, talking through problems as I solved them. The value wasn't production quality. The value was information and personality. People watch YouTube for companionship as much as education. I was the grumpy old guy in New Jersey figuring out AI filmmaking in real time. That was the show.

TikTok and YouTube Shorts are the front porch. That's where quick hits go. Fifteen to sixty seconds. The satisfying moment when a bad generation suddenly becomes a good one. A before-and-after of a character design. A clip of the film itself with text overlay. I posted a Short of the lead character's final avatar rotating slowly while I listed the prompt details in text on screen. It got eighty thousand views. Eighty thousand. For a text prompt breakdown. The algorithm doesn't care about your budget or your pedigree. It cares about engagement. And people engage with process content because it feels authentic. Because it teaches them something. Because they can imagine themselves doing it.

Instagram is the photo album. Still images, story posts, carousel posts showing iteration — generation one looking weird, generation two getting closer, generation three nailing it. I posted a carousel of a supporting character's kitchen: the first attempt had walls that looked like a dentist's office, the second had a window in the wrong place, the third had the yellow walls and wooden table that made it into the film. The caption was just my iteration notes. People swiped through it like a detective novel. Each slide was a clue in how AI image generation actually works.

Twitter — I'm not calling it X, I'm sixty-five — is the bar where filmmakers hang out. That's where you share wins, complain about tools, make jokes about your own failures, and engage with other creators' work. I spent thirty minutes every morning on there, not posting my own stuff but responding to other AI filmmakers. Commenting on their work. Sharing their posts. Asking questions. Community is a two-way street, and if you're only ever talking about yourself, you're that guy at the party who corners people and describes his screenplay for forty-five minutes. Don't be that guy. Be the person who shows up to support other people and then, occasionally, shares something of your own that benefits from the goodwill you've built.

Pro Tip: Your B-roll outtakes are content gold. Every shot that didn't make the film — every weird generation, every background that looked wrong, every character pose that had an extra finger or a face that melted slightly — is a piece of content. I made a ninety-second montage of failed generations from the sample short, set it to jaunty music, and titled it "What AI Thinks Grief Looks Like (Wrong Answers Only)." It got more views than my serious posts because it was funny and honest. People love seeing the mess. It makes them feel better about their own mess. And it makes them trust you when you show them the finished work, because they know you earned it.

Pro Tip: Post your failed generations with commentary about why they failed. Not just "lol this looks bad" but real analysis. "This prompt specified 'warm amber lighting' but Seedance interpreted that as 'the sun is exploding in this room.' The problem was I didn't specify the light source. When I added 'from a desk lamp on the left' in the next version, it worked." That's valuable to other AI filmmakers. That's why they follow you. Not because you're perfect but because you're figuring it out out loud and saving them time.

The 90/10 Rule

I learned this the hard way, like I learn everything. My first attempts at audience building were just promotional posts. "Watch my new film." "Link in bio." "Just dropped my latest project." Crickets. Absolute silence. Because nobody follows an account that only asks for things. Imagine having a friend who only calls you when he needs help moving. You stop answering. But a friend who calls to tell you about a great diner he found, who sends you funny photos of his dog, who asks how your mother's doing, and occasionally mentions he's trying to get rid of a couch — that friend gets his couch moved.

I live by what I call the 90/10 rule. Ninety percent of what I post is value or entertainment for the audience. Only ten percent is promotion of my own work. That ninety percent includes behind-the-scenes process, tips I've learned, funny observations about AI tools, shoutouts to other creators' work, honest reviews of software, dumb jokes about filmmaking. The ten percent is the ask — watch this, share this, here's the link. And because I've spent ninety percent of my time giving people something, the ten percent where I ask is received with generosity instead of resentment. People want to support creators who've supported them. That's not strategy. That's human nature. Even Frank understands that, and Frank once tried to fix his own roof.

The 90/10 rule played out perfectly when the sample short was finally released. I made one announcement post — one — across all platforms. "It's done. Twelve minutes. Made with AI tools in my spare time over four months. Here's the link." That post, combined with four months of consistent value, drove over ten thousand views in the first week. Not because the announcement was magic. Because two thousand people had been waiting for it. They'd watched it grow. They felt like they'd helped, because in small ways, they had. That wedding ring suggestion from a commenter made it into the final design. That ambient drone track recommendation from a Discord thread became the trailer music. They were invested. And invested people share.

Building What You Actually Own

Here's something the platforms don't want you to think about. You don't own your YouTube channel. You don't own your TikTok account. You don't own your Twitter following. You're building a house on someone else's land, and they can change the rules, change the algorithm, or delete your account any time they want. Ask anyone who built a massive Facebook page in 2012 how that worked out for them. Go ahead. I'll wait.

The only platform you truly own is your email list. It is the most unsexy, most important tool in your entire distribution strategy. An email list is yours. Nobody can take it away. Nobody can change an algorithm and hide your emails from the people who signed up to receive them. The open rate for email is still — still, after all these years — higher than the organic reach of any social post on any platform. When I email my list about a new film, forty percent of them open it. When I post about the same film on social media, maybe five percent of my followers see it. Those are not opinions. Those are numbers.

I started building my email list during the sample short by offering something simple: a PDF of my complete prompt library. Every prompt I'd used to generate characters, environments, B-roll. Organized by category. With notes about what worked and what didn't. I made a landing page in about an hour using a free tool, put the link in my bio on every platform, and mentioned it in every YouTube video. "If you want the full prompt list I'm using for this film, link below. Free. Just your email so I can let you know when the film's out." I got four hundred email subscribers in the first month. Four hundred people who raised their hands and said "I want to hear from this grumpy old man again."

That list is now over fifteen hundred people. When I released the second sample short — a four-minute comedy I made for thirty bucks — I sent one email. Subject line: "I made something stupid." The email had a screenshot of Harold in his disheveled undershirt on church steps, two sentences about the premise, and a link. Forty-two percent open rate. Hundreds of clicks. For a thirty-dollar film. That email list will be with me for every project I make for the rest of my life. Social platforms will come and go. Algorithms will shift. My email list is mine. Start building it today. Not tomorrow. Not when your film is done. Today. Offer something valuable in exchange for an email address. A prompt library. A behind-the-scenes video.

A five-minute consultation where you review someone's AI film idea. Give first. Collect second. Use forever.

Pro Tip: Create countdown posts in the final week before release. I posted a "7 days" image — a still from the film with text overlay — then "6 days" with a different still, all the way down. Each one got more engagement than the last because anticipation builds on itself. The day before release, I posted a clip of the trailer with the caption "Tomorrow." Nothing else. Just "Tomorrow." That was my most-liked post of the entire campaign. Sometimes saying less is everything. The silence makes people lean in.

Common Mistake: Waiting until your film is finished to start building audience. This is the Frank mistake. This is the note-on-the-door mistake. If you read nothing else in this chapter, read this: start documenting your process before you have anything finished to show. Start today. Open HeyGen, generate one character, screen-record the process, post it with a casual caption about what you're trying to make. That's your first piece of content. That's the beginning of your audience. Everything else compounds from there. But you have to start. The filmmaker who waits for perfection before showing up publicly is the filmmaker who premieres to an empty room. I've been that filmmaker. It hurts. Don't be that filmmaker.

Common Mistake: Treating social media like a billboard instead of a conversation. Nobody wants to follow a sign. They want to follow a person. Reply to your comments. Ask questions in your posts. "What should the lead character have on his desk besides the mixing board? I'm stuck." People love being asked. They love having input. They love feeling like insiders in a creative process. Every comment you ignore is a person who decided not to engage next time. Every question you answer is a relationship you're building. It takes time. It takes consistency. It takes genuinely caring about the people who show up for you. There's no shortcut. But there's also no substitute.

Frank's barbecue would have been full if he'd called people, asked what they wanted him to cook, made them feel like it was their party too.

Common Mistake: Buying followers. I can't believe I have to say this, but I see it constantly. Someone has three hundred real followers, gets impatient, buys five thousand fake ones, and thinks they've accomplished something. Those fake followers will never watch your film. They will never share your work. They will never buy anything from you. They are numbers on a screen that mean less than nothing because they actively hurt your reach — platforms see low engagement from your followers and stop showing your content to your real ones. Buying followers is like paying people to stand in your front yard at a barbecue so it looks crowded, except the people are cardboard cutouts and the real guests can't get through the door. Don't do it.

It's embarrassing. And worse — it's useless.

The Possibilities Nobody's Talking About

Now let me get excited again. Because audience building, when you really think about what's possible, stops being a chore and starts being the most creative part of the whole process.

What if you built an entire course or community around teaching your filmmaking process? Not a cheesy "make six figures with AI" course — I'm talking about the real thing. A group of filmmakers at your level, learning together, sharing prompts, critiquing each other's work. I started a Discord server during the sample short with no plan beyond "maybe people will want to talk about this stuff." It now has eight hundred members. Eight hundred filmmakers, hobbyists, curious people who show up every day to share what they're making and help each other solve problems. I spend an hour a day in there, answering questions, sharing work-in-progress, learning things I didn't know from people half my age. That community is not a marketing channel.

It's become my film school, my writer's room, my support system. And yes — when I release something, they care. Because I'm one of them, not someone trying to sell them something.

What if you livestreamed your generation sessions? Not edited highlights — raw, real-time, mistakes and all. I tried this on a whim one Saturday morning. Streamed myself generating environments for the second sample short. The funeral home interior. Took me twelve tries to get it right — the first attempts looked like a hotel lobby, then a dentist's office, then something that can only be described as a spaceship designed by someone who had only heard funerals described to them. Chat was going crazy with suggestions. "Add 'dark mahogany paneling' to the prompt." "Try specifying 1940s architecture." "Less light! Funeral homes are dim!" I tried their suggestions live. Some worked. Some failed spectacularly. The stream lasted two hours.

Peak concurrent viewers: eighty-three people. That's not viral numbers. But eighty-three people spent two hours of their Saturday watching me fail and iterate and eventually succeed. At the end, I exported the final funeral home environment and held it up to camera like a trophy. Chat erupted. Those eighty-three people felt like they'd been in the room with me. And they were the first ones to watch the second sample short when it came out. Because it was partly theirs.

And here's the wild one. What if your audience voted on creative decisions? I'm not saying let them write your film — it's your film, not a committee project. But what if you gave them real choices and honored the results? I did a poll during the sample short about the absent wife's voice. Two options. Voice A was warmer, more melodic, like a woman who sang lullabies. Voice B was thinner, slightly cracked, like someone who had been ill. I posted samples of both. No context about which way I was leaning. Just: "Which voice feels right for a woman who's gone but still present in recordings?" Voice B won by a landslide — seventy-eight percent. I'd been leaning toward Voice A. I went with Voice B. And they were right. The film was better for it.

My audience, those two thousand people who'd been following along, made a creative choice that improved the final work. That's not crowd-sourcing. That's collaboration. That's something traditional filmmaking literally cannot do at scale. Spielberg can't poll a million people about which take to use. But I can poll two thousand about which voice fits. And I did. And it worked.

The implications of this are staggering if you sit with them. What if your next film had a "community edition" where audience suggestions shaped the final cut? What if you released two versions — your cut and the audience's cut — and let people compare? What if your audience became your co-writers, not by replacing your vision but by stress-testing it, by offering perspectives you didn't have, by catching things you missed because you're too close to the work? This is what AI filmmaking enables that nobody is talking about. Not just faster production. Not just cheaper tools. The ability to make films with your audience instead of for them. The ability to turn distribution from a broadcast into a conversation. From a megaphone into a dinner party.

Frank came over last week. He was standing in my driveway holding a plate of ribs — practice, he said, for the Fourth of July. He'd been reading my posts about The sample short's release. "So let me get this straight," he said, mouth half full of rib meat. "You told people about your movie while you were making it, and then when it was done, they wanted to watch it?" I nodded. "That's it?" He seemed almost offended by the simplicity. "That's the whole secret?" I told him it wasn't a secret. It was just work. Consistent, daily, sometimes boring work. Showing up. Giving more than you ask. Making people feel like they're part of something instead of just consumers of something. Frank chewed on that — the concept, not the rib, though he did that too.

"Huh," he said finally. "So if I'd called people and asked them what they wanted to eat, and maybe had them bring a side dish, and like, posted pictures of the ribs while I was cooking them..." He trailed off. "Your yard would've been full," I said. He nodded. Walked back to his house with his empty plate. I saw him on his phone before he even got through his front door. Probably texting. Probably starting early this time. Frank learns slow, but he learns. And so do I. So do all of us. We just have to start before we're ready, share before we're finished, and trust that the audience we build with generosity will be there when we finally have something to give them.

The film on your drive is not a film yet. It's a file. It only becomes a film when someone watches it. And the path from file to audience doesn't start on release day. It starts today, with whatever you're working on right now, and the decision to let people in on the secret. To make the making of it as interesting as the thing itself. To understand that in the age of AI filmmaking, your process is your product until your product is ready — and even then, the process was what made people care. I learned this after four films that nobody watched and one film that people did. The difference was not the quality of the work. The difference was whether anyone knew the work existed. Fix that. Start now. Post something.

And then post something again tomorrow. Ribs don't eat themselves. Films don't watch themselves. And audiences don't build themselves. That's your job. And honestly? It's almost as fun as making the thing in the first place. Almost.

Chapter 15: Festivals, Screens, and the Smell of Wet Popcorn

The carpet at the Montclair Film Festival smelled like wet popcorn and anxiety. I remember this because I was staring at it, trying to convince my stomach that vomit was not an appropriate response to art. I'd been to festivals before as an attendee — sitting in the back row, sneaking a flask of coffee, judging other people's movies the way Neighbor Frank judges everyone's lawn. But this was the first time I'd walked into a theater as a filmmaker with an AI-made film in the program. My film. the sample short. Twelve minutes of grief and audio frequencies and a dead woman talking through a computer-generated avatar, and I was about to watch it with two hundred strangers who had no idea what they were in for.

I found a seat in the fifth row, which is where I always sit because the front makes your neck hurt and the back feels like you're hiding. The lights were still up. People were shuffling in with their plastic cups of wine and their expectations. A couple of twenty-somethings sat in front of me, talking about their own short film that had screened earlier. "We shot on 16mm," the guy said, like that was a moral achievement. His girlfriend nodded solemnly, as if he'd just confessed to feeding the hungry. I kept my mouth shut. I was not there to start a fight about film stock versus algorithms.

I was there because I had spent $847 making a movie in my basement and some programmer had decided it was good enough to show in a real theater with real seats that squeaked when you shifted your weight.

The lights went down. The festival bumper played — a slick montage of sponsor logos that probably cost more than my entire production budget. Then the title card: "the sample short, by Jim Bock." Twelve minutes. My hands were sweating. I could feel my pulse in my fingertips. On screen, the lead character appeared in his navy sweater with the leather elbow patches, sitting in his home studio surrounded by recording equipment that I'd described to an AI image generator and then animated with tools that hadn't existed five years earlier. The audience was quiet. Not the polite quiet of people waiting for something to happen. The real quiet of people being pulled into something.

Then the absent wife appeared. the lead character's dead wife, rendered as a Photo Avatar through HeyGen, her voice warm and present through ElevenLabs. I heard someone in the row behind me whisper, "Is that real?" Not "is that good" — "is that real." That's the question AI filmmaking always gets first. Not whether it works, but whether it's real. I wanted to turn around and say, "Real enough to make you forget about your wine, apparently." But I didn't. I sat there and watched my movie and felt something I hadn't expected to feel: pride. Not defensiveness. Not the preemptive shame of someone trying to justify their methods. Just the simple, stubborn pride of a guy from New Jersey who made something that worked.

When the lights came up, the Q&A started. A woman in the third row asked the question I'd been dreading: "How much of this was AI?" I told her the truth. The avatars were AI-generated. The voices were AI-generated through ElevenLabs — the lead character's voice was a Professional Voice Clone trained on recordings from a voice actor friend, and the absent wife used a voice from the platform's library. The B-roll came from Seedance. I edited the whole thing in DaVinci Resolve, which is free, by the way, and if Blackmagic Design ever starts charging for it I'm moving to Canada. The audience didn't boo. They didn't throw their wine cups. They asked more questions. How did you get the voice to sound emotional? How did you match the lighting between the AI avatars?

One guy wanted to know if he could do the same thing for his documentary about competitive dog grooming. I told him yes, absolutely, and good luck with the dogs.

That was my first festival experience with an AI film. It wasn't the last. But it taught me something that I've carried through every submission since: festivals don't care how you made your movie. They care whether your movie makes an audience feel something. The programmers, the judges, the people sitting in those squeaky seats — they're not running a technology audit. They're looking for stories that stick. If your AI film tells a story that sticks, you're in the game. If it doesn't, no amount of technical transparency is going to save you.

The Festival World Is a Mess Right Now, and That's Good for You

Here's the thing about the festival circuit in 2024 and beyond: nobody knows what the hell is going on. And I mean nobody. I've talked to programmers who are actively excited about AI filmmaking because it brings them weird, unexpected work from people who couldn't afford to make films before. I've talked to programmers who think AI is the devil's photocopier and want nothing to do with it. I've talked to programmers who are just confused — they got a film that says "AI-assisted" in the submission notes and they have no idea what that means or whether they should care. The whole system is in flux. Flux is opportunity.

When I started submitting the sample short, I did what I always do: I asked Chuck the plumber. Chuck was fixing my kitchen sink at the time, elbow-deep in pipes that hadn't been cleaned since the Clinton administration, and I said, "Chuck, you ever submit a film to a festival?" He looked at me like I'd asked if he'd ever been to the moon. Then he said something I've thought about ever since: "Jim, you know what the difference is between a good pipe and a bad pipe? The good one moves water. That's it. Everything else is marketing." Chuck had never made a film in his life, but he'd nailed the essential truth. Festivals are pipes. They move your film from your hard drive to an audience. Some pipes are fancy and have gold-plated fittings.

Some are PVC in your basement. But the question isn't which pipe is prettier. The question is whether your water flows through it.

So let's talk about which pipes are currently open for AI filmmakers. Tech-forward festivals are your best bet. Events like the Silicon Valley Asian Film Festival, SXSW's experimental programming, and various digital arts festivals have been openly receptive to AI work. They're already thinking about technology as a creative tool, not a threat. Experimental showcases are gold mines — places like the Ann Arbor Film Festival, the Images Festival in Toronto, or any festival with an "experimental" or "new media" category. Those programmers wake up every morning hoping someone sends them something they've never seen before. A well-made AI film is exactly that.

Digital art festivals, especially ones connected to museums or universities, are increasingly programming short films alongside interactive installations and generative art. Your AI film is not weird to them. It's Tuesday.

Then there are the genre festivals. the sample short is a thriller with grief baked into its bones, so I targeted festivals with strong thriller or psychological drama programming. Genre festivals — horror fests, sci-fi fests, mystery fests — tend to care more about whether your film delivers the goods than about your production methods. If your AI horror short makes people jump, the programmer is not going to file a complaint about your process. There are also a growing number of festivals specifically dedicated to AI-generated or AI-assisted work. These are popping up like mushrooms after rain, and while they're not all well-funded or well-attended yet, they're hungry for content.

Being in the first wave of a new festival means you're a big fish in a small pond, which is a lot more fun than being plankton in the ocean.

Disclosure: The Honesty Game

This is where people get themselves twisted into knots. Should you disclose that you used AI tools? How much should you disclose? What if the festival doesn't have a policy? What if they do have a policy and it's written by someone who thinks ChatGPT is a person?

My rule is simple: disclose what they ask for, be honest when asked directly, and don't volunteer your entire tool chain like you're reading a grocery receipt. If a festival submission form has a field about production methods, fill it out truthfully. "AI-assisted production" is usually enough. If they want more detail, give it to them. But don't write a dissertation. No programmer has ever said, "I was going to accept this film, but the director's statement about his render pipeline was only two paragraphs, so I passed."

When the sample short played at its three festivals — Montclair, the New Jersey Film Festival in New Brunswick, and the Garden State Film Festival in Atlantic City — the disclosure conversation went three different ways. At Montclair, I put "AI-assisted" on the submission form and nobody mentioned it again until the Q&A. At New Jersey Film Festival, the programmer called me before the screening and asked specific questions: which tools, how much of the film was AI versus traditional editing, whether the music was AI-generated. I answered honestly. The music was from Suno and Udio, heavily edited in post. The avatars were HeyGen. The voices were ElevenLabs with custom PVC voice IDs. He said okay, thanks for being upfront, and programmed it anyway.

At Garden State, they didn't ask at all. I showed up, they screened it, and the only question I got was about the emotional ending.

Here's my honest take: disclosure matters because integrity matters. But disclosure is not the same as apology. You're not confessing to a crime. You're describing your tools. If a carpenter showed up to build your deck and you asked what tools he used, and he said "laser-guided miter saw," you wouldn't say "that's cheating, I wanted you to use a handsaw." You'd say "great, as long as the deck doesn't collapse." The deck is the movie. If it holds up, the tools are irrelevant.

That said, some festivals are actively hostile to AI work, and you need to know who they are. Read their submission guidelines carefully. If a festival says "no AI-generated content" or "must be shot on traditional media," respect that. Don't try to sneak your AI film into a festival that has explicitly banned it. That's not clever; that's dishonest, and it hurts all of us who are trying to build a legitimate practice around these tools. The goal is to make AI filmmaking normal, not to make AI filmmakers seem deceptive.

The Submission Strategy: Don't Spray and Pray

I know a guy — let's call him Dave because his name is Dave — who submitted his short film to eighty-seven festivals. Eighty-seven. He spent $4,200 on submission fees, which is more than the budget of the sample short and the second sample short combined. He got into four festivals. Four out of eighty-seven. That's not a strategy; that's a gambling addiction with better stationery.

Don't be Dave. Target ten to fifteen festivals that actually fit your film. Here's how you choose them. Start with geography. Regional festivals love local filmmakers. If you're in New Jersey, submit to New Jersey festivals first. They want to program local talent because it sells tickets and gets local press. the sample short screened at three festivals, and all three were within driving distance of my house. I could show up for the screening, do the Q&A, shake hands with the programmers, and be home in time to disappoint my wife by forgetting to take out the trash. Regional festivals are your foundation.

Then look at genre fit. If you made a comedy, submit to comedy festivals. If you made a horror film, submit to horror festivals. If you made a twelve-minute psychological thriller about grief and audio frequencies, submit to festivals that program thrillers and dramas. This sounds obvious, but I cannot tell you how many filmmakers ignore it. They submit their experimental AI short to a festival that only programs documentaries about social justice, and then they're shocked when they get rejected. The rejection isn't about your film being bad. It's about your film being wrong for that festival. That's like bringing sushi to a pizza party. The sushi might be excellent. Nobody at the pizza party cares.

Length matters too. the sample short is twelve minutes. That puts it in the sweet spot for most short film programs, which typically want films between five and twenty minutes. the second sample short is four minutes, which is actually harder to program because it's too short for a main slot and too long for a bumper. Festivals have limited programming time and they want to fill it efficiently. A four-minute film means they need to find three more short films to fill an hour. A twelve-minute film is a nice, tidy chunk of programming that leaves room for two or three other films. If your film is under five minutes, target festivals with specific "micro-short" or "short short" categories. They exist, and they're designed for exactly your situation.

The Submission Package: Making Your Film Look Professional

Every festival submission needs a package. This is not optional. You need a synopsis, a director's statement, production stills, a press kit, and a cast and crew list. For an AI film, the cast and crew list is where things get interesting. Who do you list? Do you credit the AI tools? Do you list yourself as the entire crew? Do you make up fake names so it looks like you had a team?

Here's what I did for the sample short. I listed myself as writer, director, producer, and editor because I was all of those things. For cast, I listed the character names with performer credits noting they were "AI-generated avatars using HeyGen technology." For voice cast, I listed the ElevenLabs voice IDs — VN_ZkPJ8sjn for the lead character, and a supporting character's voice model with its warm but impatient edge. I didn't list twenty AI tools in the credits. I didn't write "special thanks to the following 47 applications." I treated the AI tools the way I'd treat any other production equipment: necessary, unglamorous, not the point.

The director's statement is where a lot of AI filmmakers go wrong. They write five hundred words about the technology — which models they used, which prompts they engineered, how many iterations it took to get the lighting right. Nobody cares. The programmer is reading your director's statement to understand what your film is about and why you made it. Write about the story. Write about the characters. Write about why the lead character's grief mattered to you, not about your Seedance render settings. If someone wants to know your technical process, they'll ask at the Q&A, and then you can talk for an hour about prompt engineering while the audience slowly loses consciousness.

Your stills matter more than you think. Most festivals use production stills for their program books, websites, and social media. If your stills look amateur, your film looks amateur before anyone presses play. Pull high-resolution frames from your film. Pick moments that are visually striking and emotionally representative. For the sample short, I used the shot of the lead character alone in his studio — the warm amber light broken only by the blue glow of the waveform monitors behind him — and the shot of a supporting character in her green cardigan reaching out toward the avatar of the absent wife. Those two images told the story better than any synopsis.

Platforms, Fees, and the Money Reality

I submitted the sample short through FilmFreeway, which is the dominant platform for festival submissions these days. Withoutabox still exists in some form — it used to be the only game in town, back when Amazon owned it and it looked like a website designed by someone who hated filmmakers. FilmFreeway is cleaner, easier to use, and most festivals are on it now. Some festivals still take direct submissions through their own websites, especially smaller regional ones. Use whatever the festival prefers. Don't be the guy who emails the programmer directly asking for a waiver because you "don't believe in submission platforms." That's not principled. That's cheap.

Submission fees add up fast. A typical festival charges between $40 and $100 for a short film submission. Early bird deadlines are cheaper, so plan ahead. the sample short spent about $400 on submissions across ten targeted festivals, which got us three acceptances. That's a 30% acceptance rate, which is actually excellent. Most filmmakers see acceptance rates between 5% and 15%. I attribute our success to two things: the film was genuinely good, and we targeted festivals that were a good fit for its genre and length. The $400 in submission fees was money well spent.

It got us three festival screenings, audience Q&As, local press coverage, and enough credibility to pitch the film to clients who ended up paying ten times the film's budget for similar work.

If you can't afford submission fees, look for festivals that offer waivers. Many festivals waive fees for local filmmakers, students, or films in specific categories. Some festivals have "no fee" submission periods. Don't be ashamed to ask. The worst they can say is no, and then you're exactly where you started, which is not a loss.

The Emotional Truth About Rejection

Let me tell you about the rejections. the sample short was rejected by seven festivals. Seven nice, polite, form-letter rejections that all said basically the same thing: "Thank you for your submission. We received a record number of entries this year and were unable to program your film." The first one stung. The second one stung a little less. By the fifth one, I was treating them like weather reports. Oh, it's raining again? Guess I'll bring an umbrella.

Rejection is the normal state of festival submission. Most films get rejected from most festivals. This is not a reflection of your worth as a filmmaker. It's a reflection of limited programming slots and subjective taste. I've seen terrible films get into major festivals because the programmer had a personal connection to the director. I've seen brilliant films get rejected because the programmer watched them on a Tuesday afternoon while recovering from food poisoning and wasn't in the mood. Festivals are run by humans, humans are inconsistent, and the sooner you accept that, the happier you'll be.

My wife has a system for dealing with festival rejections. Every time we get one, she makes me say one thing I like about the film out loud. Not something the programmer said, not something a reviewer said. Something I like. "I like the way the absent wife's voice catches when she says the lead character's name." "I like the blue light in the studio scene." "I like that we made this for $847 and it still makes people cry." It sounds corny, but it works. The rejections stop being verdicts and start being footnotes.

Pro Tip: Submit to festivals in your home region first, not last. Local festivals want local filmmakers. They want to be able to say they discovered you before you got famous. They want local press coverage. They want you to show up for the screening and bring your friends. The sample short's first acceptance was the New Jersey Film Festival, and that screening led directly to the programmer at Garden State hearing about the film and seeking us out. One local festival led to another, which led to credibility, which led to everything else.

Pro Tip: Use festival deadlines as motivation to finish your film. There's nothing like a submission deadline two weeks away to cure perfectionism. When I was editing the sample short, I set my first festival deadline as my hard finish date. The film wasn't perfect — the color grade was a little warm in the second scene, and I never fully fixed the audio transition between the lead character's studio and the flashback — but it was done, and done was better than perfect. Perfect is the enemy of screened.

Pro Tip: The "short film plus Q&A" format is a gift to AI filmmakers. Audiences at festival screenings are curious about how you made your film. They want to know your process. If you made your film with AI tools, that Q&A becomes a mini-tutorial. People will ask questions. They'll take notes. They'll approach you afterward and want to talk for twenty minutes about HeyGen avatars or ElevenLabs voice cloning. This is networking without the business cards. Embrace it.

Common Mistake: Submitting your film before it's actually finished. I know you're excited. I know you want to get it out there. But if your film has placeholder audio, rough color, or a temp score, don't submit it. Programmers can tell. They watch hundreds of films and they know what unfinished looks like. Wait until it's done. The festival will be there next year. An unfinished submission lasts forever in someone's memory as "that rough cut that wasn't ready."

Common Mistake: Writing a director's statement about the technology instead of the story. I covered this above, but it bears repeating because it's the single most common mistake I see from AI filmmakers. Your director's statement is not a product review. It's not a technical manual. It's a story about why you made this film and what it means to you. Write about the lead character's grief. Write about what it felt like to hear the absent wife's voice come through the speakers for the first time. Write about why this story needed to be told. The technology is how you told it. The story is why.

Common Mistake: Listing twenty AI tools in your credits. This makes your film look like a tech demo, not a movie. List the essential tools. List yourself as the creator. Treat AI the way you'd treat a camera or editing software — as a production tool, not a creative partner. You are the filmmaker. The AI is your equipment. Don't give your equipment top billing.

What If You Went Wild With This?

Here's where I let my imagination run loose, because the festival world is changing so fast that the conventional advice I'm giving you might be obsolete in two years. What if you started your own festival specifically for AI films? I'm not talking about a big, expensive, red-carpet affair. I'm talking about a one-night event in your local community theater or university auditorium. Ten AI short films. A Q&A with the filmmakers. A reception afterward where everyone argues about whether AI art is real art until someone eats too much cheese and the conversation devolves into personal anecdotes. You could do this for under $1,000 if you found a free venue and asked the filmmakers to promote it themselves. You'd be building community.

You'd be creating a market where none existed. You'd be a festival director, which looks great on your bio and gives you an excuse to wear a blazer.

What if festivals become irrelevant? This is the possibility that keeps me up at night, in a good way. YouTube has two billion users. A successful festival screening reaches two hundred people, if you're lucky. What if the real premiere of your AI film isn't in a theater at all? What if it's a YouTube upload at 9 AM on a Tuesday, and by Wednesday morning it's been seen by fifty thousand people, and by Friday it's being shared in Reddit threads and Discord servers and someone in Finland has made a reaction video? The romantic in me loves the theater. The pragmatist in me knows that reach matters more than prestige. I'm not saying abandon festivals. I'm saying don't let festivals be your only path. They're one pipe. There are others.

And here's the wildest thought, the one that makes me grin like an idiot when I'm supposed to be editing: what if your AI film wins an Oscar? Not today, not with the current rules, which require theatrical exhibition and various other hoops. But the rules change. The Academy added a Best Animated Feature category in 2001 because animated films had become too important to ignore. They expanded the Best Picture category from five to ten nominees because they were leaving out popular films that audiences actually cared about. At some point — five years from now, ten years from now — there will be an AI-generated film so good, so undeniable, that the Academy has to reckon with it. Why shouldn't it be yours? You have the tools. You have the story.

You're reading this book, which means you're already ahead of ninety percent of the people who talk about making films but never do. Someone is going to be the first AI filmmaker at the Oscars. Why not you?

I think about that sometimes when I'm exporting a render at 3 AM and the progress bar is moving so slowly I wonder if time has stopped. I think about the possibility that this weird, grumpy, basement-based way of making films might actually be the future. Not a diminished future. Not a cheapened future. A future where anyone with a story and a laptop and enough stubbornness can make something that matters. That's worth staying up for. That's worth the rejection letters and the submission fees and the nervous stomach in the fifth row.

The next chapter is about what happens after the festival. You screened your film. People saw it. Now what? Now you get it in front of the biggest audience possible, and you don't need a programmer's permission to do it. Let's talk about distribution.

Chapter 16: Distribution — Getting Your Film Off Your Hard Drive

I posted the sample short to YouTube at 11:47 PM on a Tuesday night in October. I remember the exact time because I had been telling myself I'd upload it "tomorrow" for three weeks, and that night I finally ran out of excuses. The color grade was acceptable. The audio was clean enough. My wife had watched the final cut and said, "It's good, Jim," which is her highest praise and also the thing she says right before asking me to finally clean the garage. I exported the file, logged into YouTube, and stared at the upload button like it was the launch control for a nuclear missile.

One click and my film would be out in the world, available to anyone with an internet connection and a tolerance for twelve-minute indie films about dead wives and sound engineering.

I clicked it. The upload bar crawled across the screen. I paced my office, which is really just the corner of my basement that doesn't have the water heater in it. When the processing was done, I made the video public. No premiere. No countdown. No fancy announcement. Just a YouTube video, live on the internet, sitting there like a puppy at the pound waiting for someone to notice it. Then I went to bed because I'm sixty-five and midnight is my hard stop.

The next morning I woke up and checked my phone before I brushed my teeth. Thirty-seven views. Thirty-seven people had watched my film while I was asleep. In my fantasy, I would wake up to ten thousand views and a comment from a major studio offering me a three-picture deal. In reality, I got thirty-seven views and one comment from a guy named "FilmBuff99" that said, "Is this AI? The faces look weird." I stared at that comment for a long time. Then I made coffee and reminded myself that every film ever uploaded to the internet started with a comment from someone saying something looked weird.

Here's the truth about distribution that nobody wants to tell you: putting your film online is easy. Getting people to watch it is hard. The upload button is the smallest part of the equation. The real work is everything that comes after — the promotion, the placement, the platform strategy, the relentless grinding of showing your film to people who have a million other things to watch. But here's the other truth: it's never been easier to reach an audience without someone's permission. You don't need a distributor. You don't need a sales agent. You don't need a festival programmer to bless your work. You need a plan, a platform, and the willingness to do the work that most filmmakers skip because they think uploading is enough.

YouTube: The Default, The Giant, The Mess

YouTube is where most people start, and for good reason. It's free. It has two billion logged-in monthly users. It handles all the bandwidth, all the formatting, all the compatibility across devices. You upload a file and suddenly anyone on Earth can watch your film on their phone, their TV, their refrigerator screen if they're fancy. That's power. That's unprecedented in human history. For most of filmmaking history, if you wanted someone in Australia to see your movie, you needed to physically ship them a reel of film. Now you click a button and a teenager in Melbourne can cry at your AI-generated scene about grief while eating Vegemite toast in bed. That's magic.

But YouTube has problems for short filmmakers. The algorithm does not love short films. It loves watch time, and a twelve-minute film has a natural ceiling. It loves engagement, and most people who watch short films don't comment or subscribe at the rate that algorithmic gods demand. It loves regular uploads, and you're a filmmaker, not a daily vlogger. You made a film. You spent months on it. You're not going to upload a new one tomorrow. The algorithm will punish you for this. The algorithm wants frequency, and you are offering art. Art is not on the algorithm's schedule.

The ad revenue is another joke. If the sample short got a hundred thousand views on YouTube, I might make enough money to buy a nice dinner. Not a fancy dinner. A nice dinner. Maybe at Olive Garden if I don't order wine. YouTube pays creators through the Partner Program, but the rates for short films are abysmal because advertisers don't want to place their ads in front of twelve-minute dramas about dead wives. They want preroll on makeup tutorials and video game streams. Your beautifully crafted short film is not a cash cow. It's a calling card. Treat it that way.

What YouTube is great for is discovery. It's the world's second-largest search engine after Google. If someone searches for "AI short film" or "indie thriller" or even just types in the title because they heard about it somewhere, YouTube will serve them your film. It's portfolio visibility. It's proof that you exist as a filmmaker. When I pitch client work now, I send them a link to the sample short on YouTube, and they can watch it immediately without downloading anything or signing up for anything. That instant accessibility is worth more than ad revenue. The film lives on YouTube not to make money but to make relationships.

Vimeo: The Filmmaker's Sanctuary

Vimeo is where I put The sample short's "official" version — the one I link in my portfolio, the one I send to festivals for consideration, the one I show clients who I want to impress. Here's why. Vimeo has better video compression than YouTube, which means your film looks sharper and cleaner. Vimeo doesn't put ads before your film, which means your audience doesn't have to watch a mattress commercial before they experience your emotional opening scene. Vimeo's player is cleaner, more professional, and customizable. You can replace the video file without changing the URL, which is a godsend when you spot a color issue after you've already sent the link to seventeen people.

You can password-protect videos, which is essential for festival submissions that haven't premiered yet. You can add a custom thumbnail that doesn't look like it was designed by a confused robot.

Vimeo costs money — the Pro plan is something like $20 a month — but it's worth it if you're serious about presenting your work professionally. I think of Vimeo as my suit and tie, and YouTube as my comfortable pants. When I'm meeting a potential client or submitting to a festival, I wear the suit. When I want maximum reach and don't care about the ad interruption, I wear the comfortable pants. Both have their place. The mistake is thinking YouTube is enough, or thinking Vimeo is the only option. Use both strategically. the sample short lives on YouTube for public discovery and on Vimeo for professional presentation.

the second sample short, being a four-minute comedy, lives only on YouTube because it's more of a casual watch, something you stumble into and laugh at and share with a friend. Different films, different platform strategies. There's no one-size-fits-all answer, and anyone who tells you there is hasn't made enough films to know better.

The Streaming Aggregators: Tubi, Plex, and the Long Tail

Beyond YouTube and Vimeo, there are streaming platforms that accept short films through aggregation services. Tubi, Plex, Freevee — these are ad-supported platforms that reach millions of viewers and will take short films if you go through the right channels. Usually that means working with an aggregator like FilmHub or BitMax, which takes a cut of revenue in exchange for placing your film on dozens of platforms.

Grumpy Old Man Tool Note: Tubi, Plex, Freevee, FilmHub, and BitMax

Tubi, Plex, Freevee, FilmHub, and BitMax belong to the world beyond festivals and YouTube. This is where distribution starts sounding official and everybody suddenly wants deliverables, metadata, captions, artwork, runtime information, and rights clarity. In other words, grown-up paperwork. Sorry, the movie business is still a business.

The possibility is reach. Ad-supported platforms and aggregators can help a finished project travel farther than your personal audience. FilmHub and BitMax-type services can become pathways into platforms when you have the right materials and rights in order. Pair that with trailers on YouTube, a press kit on your website, an email list, and social clips, and you start building an ecosystem instead of throwing one upload into the ocean.

The limitation is expectations. Distribution does not magically create demand. A platform listing is not a parade. You still need artwork, positioning, audience building, keywords, trailer strategy, and a reason for people to care. Getting into the store is nice. Somebody still has to walk down the aisle and pick up the box.

Let me be straight with you: the revenue from these platforms for a short film is practically nothing. You're not going to retire on Tubi ad impressions. What you get is legitimacy and reach. When someone asks where they can watch your film and you say "it's on Tubi," that sounds more professional than "search for it on YouTube and scroll past the sponsored content." It's a line on your bio. It's a credential. And for some audiences — especially older viewers who don't spend time on YouTube — it's the only way they'll ever find your work.

I haven't put the sample short on Tubi yet, mostly because the twelve-minute runtime is awkward for their programming model, which is built around feature-length content. But I've seen AI short films do well on these platforms when they're packaged as collections — five AI shorts bundled into an hour-long program. That's a strategy worth considering. If you know four other AI filmmakers, you could create an anthology, submit it as a package, and suddenly everyone's film is getting exposure on a platform that individual shorts can't access alone.

Direct-to-Consumer: Your House, Your Rules

The most underrated distribution strategy in the world is the simplest one: put your film on your own website and tell people to go there. Full control. No algorithm. No ads. No platform risk. If YouTube changes its terms tomorrow and bans AI content — which they won't, but they could — your film is still safe on your own domain. If Vimeo raises its prices to $200 a month, you just stop using them. Your website is your house. Nobody can change the rules but you.

Here's what I did. I embedded the sample short on my NJ Film Productions website alongside a "how I made this" page that breaks down the process, lists the tools, and includes behind-the-scenes notes. The film plays at the top. Below it, there's text about the production, links to the tools I used, and a contact form for people who want to hire me to make something similar. That page is now my highest-converting sales tool. People watch the film, they read about how it was made, and they think "I want something like that for my business" or "I want to learn how to do this." Then they fill out the form. That single page has generated more client revenue than every festival screening combined.

The direct-to-consumer path also means owning your audience relationship. When someone watches your film on YouTube, YouTube owns that relationship. You don't get their email address. You don't know who they are. They might subscribe to your channel, but they might not, and even if they do YouTube decides whether they ever see your content again. On your own website, you can collect email addresses, build a list, notify people when your next film comes out. Email is the only distribution channel that nobody can take away from you. Not Google, not Meta, not whatever company buys whatever platform next year. Email is yours. Treat it like gold.

Release Windows: Timing Is Everything

There's an old-school concept in film distribution called "release windows" — the idea that a film premieres in one venue before it becomes available elsewhere. Theaters first, then home video, then streaming, then cable TV. Each window is designed to maximize revenue at each stage. For short films made by solo AI filmmakers, revenue isn't really the point, but the window concept still matters because of one word: premiere.

If you want to screen at festivals, most of them care about premiere status. A world premiere means your film has never been publicly screened anywhere. A North American premiere means it hasn't screened on this continent. A regional premiere means it hasn't screened in that specific region. Some festivals only accept world premieres. Some only care that it's a premiere for their specific festival. Read the rules. If you upload your film to YouTube before submitting to festivals, you might disqualify yourself from festivals that require unreleased work. the sample short followed the "festival first, then public" strategy.

We submitted to festivals, got our acceptances, screened at the three that programmed us, and then released publicly on YouTube and Vimeo after the festival run was complete. That gave us the premiere credentials without preventing us from reaching a wider audience.

The alternative is the "release everywhere simultaneously" strategy, which is what I did with the second sample short. Four-minute comedies don't benefit from festival prestige the way dramas do, and the festival circuit moves slowly. By the time the second sample short would have made it through a festival submission cycle, the joke would have been stale. So I uploaded it everywhere at once — YouTube, Vimeo, Twitter, Instagram, even TikTok, where I posted the full four minutes as a single video because TikTok's length limits had expanded. It got more views in its first week than the sample short got in its first month, partly because comedy travels faster than drama, and partly because releasing everywhere simultaneously creates a splash instead of a ripple.

Different films, different strategies. There is no universal right answer. There is only the right answer for your specific film and your specific goals.

Social Media as Distribution: The New Theaters

I want to spend a minute on something that would have sounded insane to me five years ago but now seems obvious: social media platforms are distribution channels. Not promotion channels for your distribution. Actual distribution channels where the film lives. TikTok, Instagram Reels, Twitter — these are where millions of people watch video content every day. Posting your full film on TikTok is not "giving it away for free." It's "screening it in a theater that holds a billion people."

The format matters, of course. A twelve-minute film doesn't fit TikTok's typical consumption pattern. But a four-minute comedy like the second sample short? That's perfect. I posted the second sample short on TikTok as a single video, added captions because most people watch without sound, and included a hook in the first three seconds — Harold Vance sitting in his disheveled black suit, staring into camera, with text overlay saying "my dead mentor left me a whoopee cushion." That got people's attention. The video got 340,000 views in its first week. Not because I'm famous. Not because TikTok's algorithm loves funeral director comedies. Because the format matched the platform, the hook was strong, and the film delivered on its promise.

YouTube Shorts and Instagram Reels work the same way. If you have a short film — under five minutes — consider posting the full thing directly on these platforms. Don't post a trailer linking elsewhere. Don't post a clip with "watch the full film on YouTube" text. Just post the film. Meet people where they are. The barrier between "scrolling" and "watching" is the single biggest obstacle in modern distribution. Every click you require is a wall. Every redirect is a door that closes. Post the film. Let people watch it. Trust that some percentage of them will want to know more and will find their way to your website or your other work.

The Revenue Reality: Don't Quit Your Day Job

I need to say this directly because I see too many young filmmakers — and by "young" I mean anyone under sixty, which is most of you — with unrealistic expectations about short film revenue. Most short films make no money. Zero. They lose money. You spend $847 on production, $400 on festival submissions, $240 on a year of Vimeo Pro, and you earn back maybe $30 in ad revenue and a free festival pass worth $25. The math does not work if you're looking at the film as a product to sell.

But the film is not the product. The film is the commercial for everything else you can do. the sample short was a $847 investment that returned itself within three months through client work. A business saw the film, contacted me through my website, and hired NJ Film Productions to make a similar piece for their brand. That client paid $8,000. Then another client saw that piece and hired us for $12,000. Then a film festival programmer saw the second client piece and invited me to speak on a panel, which led to three more inquiries. The film itself never made a dime in direct revenue. The film made everything else possible.

the second sample short cost about $30 to make and has earned about $45 in total ad revenue across all platforms. But it lives on my portfolio page as proof that I can make comedy, that AI tools can be funny, that a four-minute film about a funeral director and a whoopee cushion can actually work. That proof has value. It doesn't show up in a bank account immediately, but it compounds over time. Every film you make is a brick in a wall you're building. The wall is your career. The wall is your credibility. The wall is the thing that eventually makes people say "oh, you're Jim Bock, I saw your film" instead of "who?"

Pro Tip: Embed your film on your own website alongside a "how I made this" page. I said this above but I'm saying it again because it's the single most valuable distribution move you can make. The film gets their attention. The behind-the-scenes page builds trust. The contact form captures the relationship. You now have a three-step funnel that works while you sleep. A guy from Ohio watched the sample short at 2 AM, read my breakdown, and emailed me at 2:17 AM asking about a project. I woke up to money in my inbox. That's distribution working.

Pro Tip: Use your film as a calling card for client pitches. When I approach a potential client, I don't send a resume. I don't list my equipment. I send a link to the sample short and say "here's what I can do for $847. Imagine what I can do with your budget." That landing page with the embedded film does more work than any pitch deck ever could. The film is the pitch. The film is the proof. The film is the conversation starter that gets you in the door.

Pro Tip: Cross-post to multiple platforms but customize your descriptions for each. On YouTube, I wrote a description optimized for search: "the sample short — AI short film about grief, sound engineering, and a father's love." On Vimeo, I wrote something more personal: "My first AI film. Made in my basement. Dedicated to anyone who's ever tried to hold onto someone's voice after they're gone." On Twitter, I posted a thread with screenshots and production notes. Same film, three different contexts. Don't be lazy and paste the same text everywhere. Each platform has its own culture. Speak the language.

Common Mistake: Uploading to YouTube and expecting views without promotion. The upload button is not a magic wand. YouTube has 500 hours of video uploaded every minute. Your film is a snowflake in an avalanche. If you don't promote it — on social media, in email newsletters, in relevant communities, through outreach to bloggers and podcasters — it will sit there with thirty-seven views and a comment from FilmBuff99 about weird-looking faces. The work starts after the upload. Most filmmakers stop there. Don't be most filmmakers.

Common Mistake: Choosing a platform for prestige instead of audience fit. I've seen filmmakers put their short films on expensive, obscure platforms because the platform sounds impressive at parties. "Oh, my film is streaming on CinephileVault Ultra." Great. CinephileVault Ultra has twelve users and nine of them are the founders' mothers. YouTube has two billion users and some of them might actually watch your film. Choose the platform that reaches your audience, not the platform that impresses people at cocktail parties. If your audience is on TikTok, be on TikTok. If your audience is on YouTube, be on YouTube. Prestige without reach is vanity.

Common Mistake: Ignoring SEO and discoverability. Your film's title, description, and tags determine whether anyone ever finds it. "the sample short" is a searchable, memorable title. If I'd called it "Untitled Project 7" or "Film Final Render V3," nobody would ever find it. Use descriptive titles. Write detailed descriptions with relevant keywords. Tag your film with terms people actually search for: "AI short film," "indie thriller," "short film 2024," the names of your actors or characters. You spent months making this thing. Spend twenty minutes on the metadata.

Common Mistake: Not having a clear call-to-action after the film ends. What do you want the viewer to do next? Subscribe to your channel? Visit your website? Hire you for a project? Email you their thoughts? If you don't tell them, they won't do anything. End your film with a card that has your website. Put a link in the description. Add an end screen on YouTube with clickable links. Guide them. They're not mind readers. They just watched your movie and felt something. Capitalize on that feeling by giving them a next step. Otherwise they close the tab and forget your name before the screen goes dark.

Possibility Sparks: The Distribution Revolution Nobody's Ready For

Let me end this chapter with some wild thoughts, because the distribution landscape is changing faster than any of the "experts" can track, and the people who win are the ones who see the shifts coming before everyone else catches up.

What if you released your film as a series of TikTok chapters? Take your twelve-minute film and break it into twenty-four thirty-second clips, each ending on a cliffhanger. Post one per day. Build anticipation. Let the algorithm work for you instead of against you. By the time you post the final clip, you have a built-in audience who's invested in the story and will share the full film when you release it as a complete piece. This is serial storytelling for the attention-economy age. Dickens did it with magazines. You're doing it with vertical video. Same principle, different century.

What if the film itself is just the entry point to an immersive world? the sample short exists as a twelve-minute video, but what if it also existed as an interactive experience where viewers could explore the lead character's studio in 3D? What if they could listen to the absent wife's audio recordings on a website designed to look like the lead character's mixing board? What if there was a Spotify playlist of the music from the film, and a behind-the-scenes podcast, and a series of social media posts from the child character's perspective, as if she were a real teenager documenting her life online? The film is the center of a universe. Distribution isn't just putting the film somewhere. It's building gravity around the film that pulls people into orbit.

Grumpy Old Man Tool Note: Spotify and Audio Extensions

Spotify matters here not because every film needs a soundtrack album, but because stories can spread through audio too. A film world can become a podcast, a character diary, a behind-the-scenes audio series, a soundtrack release, or a director commentary. If the story has a voice, do not assume it only belongs on a screen.

The possibility is expansion. Use ElevenLabs for character voice experiments, Suno or Udio for temp songs, licensed music for safer public release, and Spotify-style audio channels to turn a small film into a larger world. A twelve-minute short can become interviews, music, commentary, and audience touchpoints. That is how a project stops being one file and starts becoming a brand.

The limitation is rights and quality. Do not dump low-effort audio online and call it strategy. Make it useful, entertaining, or emotionally connected to the film. And keep track of licenses. Music rights are not a place to get cute. That road has potholes big enough to swallow a Buick.

And here's the one that keeps me up past my bedtime: what if every viewer got a slightly different cut? AI filmmaking tools are advancing so quickly that personalized films are becoming possible. Imagine uploading the sample short to a platform that subtly adjusts the color grade based on the viewer's local weather. Or changes the background music based on their Spotify preferences. Or generates a slightly different ending depending on what time of day they watch. This isn't science fiction. The technology exists today. It just hasn't been productized for filmmakers yet. But it will be. And when it is, the filmmaker who figures out how to make personalized cinema will have created something that traditional filmmaking literally cannot replicate.

That is distribution as innovation. That is the future, and it's closer than you think.

The path from finished film to watched film is not a straight line. It's a maze with multiple entrances and exits. YouTube is one door. Vimeo is another. Your own website is the door you build yourself. Festivals are doors that someone else opens for you, sometimes, if they feel like it. Social media is a window you climb through when the doors are locked. The point is not to find the one right way. The point is to walk through as many doors as you can, to reach as many people as possible, to get your film in front of eyeballs by any means necessary.

Because a film that nobody sees is just data on a hard drive. And data on a hard drive doesn't change anyone's life. Data on a hard drive doesn't make a client call you. Data on a hard drive doesn't start a conversation about what AI filmmaking can become. Your film deserves to be seen. The platforms are there. The audience is waiting. All that's left is the work of getting it to them, one upload, one share, one conversation at a time. That's distribution. That's the job. And you're the only one who can do it for your film.

Chapter 17: AI Dubbing — The World Speaks Your Language

I'm sitting in the fifth row at the Montclair Film Festival, watching the sample short on the big screen, remembering the first time I saw it play for an audience. The scene where the lead character listens to the absent wife's recording through his headphones — the close-up where his eyes well up — it's hitting exactly the way I hoped. The lights come up, a few people sniffle, and then this woman in the second row turns around and asks me, completely earnestly, "Does it come in Spanish? My mother would love this, but she only speaks Spanish." And I had to tell her the truth. I said, "Not yet. Maybe someday." That was 2023.

Back then, making a Spanish version meant hiring a voice actor, booking a studio, syncing audio, re-rendering the whole thing — probably another eight hundred bucks, easy. I went home that night and told my wife about it over reheated lasagna. She said, "So learn Spanish." I said, "I'm sixty-five. I can barely work the remote." She wasn't wrong though. The barrier was real. A film that cost $847 to make would cost the same again just to speak a different language. That felt broken to me. Like buying a car and paying the sticker price again just to change the paint color.

Here's what changed. AI dubbing and lip-sync tools arrived, and they flipped the economics upside down. Now that same film — the sample short — can speak forty languages for zero additional dollars. I'm not exaggerating. The marginal cost of making my twelve-minute film accessible to a few billion more people is exactly zero. That's not a typo. That's not marketing fluff from some Silicon Valley kid in a hoodie. That's the actual math.

Let me walk you through how this works, because when I first saw it, I felt like Chuck the plumber must have felt when he discovered power tools. He still talks about the day he switched from hand-cranking pipe threaders to electric. "Jim," he told me, standing in my basement staring at my water heater like it owed him money, "I did in twenty minutes what used to take me three hours. I almost felt guilty charging the customer." I told him to keep charging. The customer doesn't care about your tools. They care about the result.

HeyGen's multilingual system works like this. You upload your film. Not a script, not an audio file — the actual video. You select your target language from a list that keeps growing. Spanish, Mandarin, Hindi, French, German, Portuguese, Japanese, Arabic, Korean, and on and on. The system analyzes the visual speech patterns, maps the mouth movements, and re-renders the video so the lips match the new language. Not perfectly — I'll get to that — but convincingly enough that most viewers won't notice unless they're looking for it. The avatar's mouth actually moves differently for Spanish than it does for English. The rhythms change. It's not just a voiceover slapped on top. It's visual dubbing.

ElevenLabs takes this a step further with what they call voice preservation in translation. This is the part that gave me goosebumps the first time I heard it. Remember the lead character's voice in the sample short? That rough, tired baritone I cloned from my own voice recordings? When I ran it through ElevenLabs' dubbing tool, the lead character didn't just start speaking Spanish with some random Spanish voice. He spoke Spanish with his voice. Same gravel, same pauses, same weight. It sounds like the same man learned Spanish overnight. I played it for my wife and she said, "That's creepy." Then she listened to the whole scene and said, "That's amazing." Both things can be true.

The tool supports something like twenty-nine languages now, and the quality varies by language pair. European languages tend to sound most natural. Asian languages with very different phonetic structures can sometimes sound slightly processed. But the gap closes every month. What sounded robotic six months ago sounds natural today.

For the sample short, I focused on three languages that made business sense. Spanish first, because the US Latino audience is enormous and underserved by indie film. Mandarin second, because the sheer population numbers make it worthwhile even if only a fraction watch. French third, because European festivals love French versions and it opens doors in Canada too. I didn't do forty languages on day one. I picked three that mattered, tested them, and let the results guide the next batch.

Now, subtitles versus dubbing. This is a creative choice, not a technical one, and AI makes both trivially easy. Some films want subtitles. Art house films, foreign cinema fans, people who believe the original performance is sacred — they want to hear the actor's real voice and read along. Subtitles preserve performance. But dubbing opens doors that subtitles can't. My mother-in-law won't watch subtitled films. Neither will millions of people who find reading along distracting or difficult. Accessibility matters. A dubbed version isn't a lesser version. It's a more version. More people can experience what you made. the sample short works beautifully with subtitles for festival audiences who appreciate that kind of thing.

But the dubbed Spanish version I sent to a community center in Union City? They played it for fifty people, mostly older women who never would have read subtitles fast enough. One of them cried. That was worth the thirty minutes it took me to generate the dub.

Here's where it gets tricky though. the second sample short taught me that comedy is harder to dub than drama. Much harder. AI translates words. It does not translate humor. When Harold delivers the limerick eulogy — "There once was a man from Nantucket" — that joke lives in English. It rhymes in English. The whoopee cushion gag works visually, so that survives. But wordplay dies in translation. I ran the second sample short through Spanish dubbing and the limerick became... fine. It conveyed the idea. But it wasn't funny. It was a man reading a poem at a funeral. The absurdity evaporated. For comedy, you have two choices. You can accept that some jokes won't land and dub anyway for the parts that work.

Or you can rewrite jokes specifically for the target language and record those as alternate versions. That's more work, but for a four-minute film, it's maybe an afternoon's effort.

Pro Tip: Plan for dubbing during your writing phase. Avoid culturally specific idioms if you know you want global distribution. "It's raining cats and dogs" doesn't survive translation. Neither does "piece of cake" or "hit the nail on the head." Write plain, specific, visual language. A character saying "I'm exhausted" travels better than "I'm running on fumes." Generate a "dubbing-friendly" script version alongside your original — just a plain-language rewrite you keep in your files. And always, always test lip sync on thirty seconds before you commit your full film. Pick a dialogue-heavy scene with lots of close-ups and run that first. It'll save you hours of re-rendering if something's off.

Common Mistake: Assuming AI translation captures nuance. It doesn't. Not yet. When the lead character says "I can't let her go" in the sample short, that line carries twelve minutes of accumulated grief. The AI translated the words correctly into Spanish, but I had a friend from Bogotá review it, and she suggested a slightly different phrasing that carried more emotional weight in Spanish. It was a small change, but it mattered. Always find one native speaker to review your dubs. Even if you don't speak the language, watch the dubbed version yourself and look for timing mismatches — moments where the mouth keeps moving after the audio stops, or where a reaction shot comes too early because the new language has different sentence lengths.

Your eye catches problems your ear can't.

What If You Went Wild With This? Here's where my imagination runs. What if you released the sample short as a "language tour" — every week for forty weeks, the film "premieres" in a new language? Each week you post the new version, share a little about the language and the culture, build an audience that follows along for the journey. What if you leaned into the synthetic quality instead of hiding it — released a deliberately stylized multilingual version where the lead character speaks a different language every sentence, the lip sync visibly shifting, creating an uncanny artistic effect that comments on global communication itself?

What if you built a toggle into your website where audiences could switch languages mid-scene, comparing how the same moment feels in Japanese versus Portuguese? The tools make all of this possible. The only limit is your willingness to experiment. Frank next door didn't understand a word of the Mandarin version I showed him, but he watched the whole thing. "I got the feeling of it," he said. "The guy misses his wife. You don't need subtitles for that." Frank's right more often than he knows.

Chapter 18: The Business That Finds You

The call came two weeks after the sample short finished its festival run. I was in my office — which is really just the spare bedroom with a desk I bought at Staples in 2011 — and my phone rang with a number I didn't recognize. I almost didn't answer. At my age, unknown numbers are usually someone trying to sell me solar panels or tell me my car warranty expired. I don't even have a car warranty. But I picked up, and this woman named Denise introduced herself as the program director for a nonprofit in Paterson that runs grief counseling workshops for families who've lost someone suddenly. She'd seen the sample short at the Montclair screening. And she asked me the question that changed how I thought about this whole thing.

"Can you make something like that for us?" Not "can we license your film." Not "can you come speak." Can you make something. For money. I stood there with the phone to my ear staring at a water stain on my ceiling that Chuck still hasn't fixed, and I realized: wait, this is a business.

I need to tell you the hard truth first, because nobody else will. Short films don't make money. Even traditionally made short films with real crews and actual budgets rarely break even. Festivals don't pay. Streaming platforms might give you a few hundred dollars if you're lucky. Theatrical release for a twelve-minute film is basically nonexistent. If you're making AI short films to get rich, you're in the wrong game. Go sell insurance. My uncle sold insurance for forty years and retired to Florida with a condo that has a pool shaped like a kidney bean. That's not happening with indie film.

But — and this is the "but" that matters — the film itself doesn't have to make money for you to make money. The skills you build, the assets you create, the reputation you establish, the proof-of-concept you now hold in your hands — those are incredibly valuable. the sample short cost $847 and returned exactly zero dollars in direct revenue. But it led to client work that paid over ten times that budget within six months. The film was a marketing expense. A loss leader. The world's most effective business card.

Let me break down what that client work actually looks like, because when I started, I had no idea what businesses even wanted. Turns out, they want a lot of things, and most of them are boring in exactly the way that makes them profitable. Training videos for companies onboarding new employees — that's a huge one. A local logistics firm paid me $2,400 to make a twenty-minute training series about warehouse safety procedures using AI avatars instead of filming actual warehouses. Think about that. Twenty minutes of professional training content for less than a traditional crew would charge for a single day of prep.

The logistics guy — his name was Rich, he talked faster than anyone I've ever met — told me they'd been quoted twelve thousand dollars by a traditional production company for the same project. I delivered in five days. He hired me for three more series within two months.

Explainer videos for software companies are another goldmine. These companies have products that are genuinely useful but impossible to describe in a sentence. They need someone to show, not tell. I made a three-minute explainer for a project management startup that showed their app in action, narrated by an AI spokesperson who walked through three use cases. Cost me about sixty dollars in platform credits. They paid $1,500 and told me it converted better than any written documentation on their website. That's the kind of feedback that gets you the next client.

Internal communications for corporations where the CEO wants to send a personal message to ten thousand employees but doesn't want to film it forty times because she's traveling or camera-shy or both. I clone her voice, create her Digital Twin once, and then she can "record" new messages monthly without ever stepping in front of a camera. One HR director told me my AI version of her CEO looked "more relaxed than the real thing." I didn't tell the CEO that.

Nonprofit storytelling — like Denise's grief counseling organization, which paid me $1,800 to make a short film about their work that they now show at every fundraiser. Denise cried when she watched the rough cut. Not because it was sad, but because it finally showed people what her organization actually did. She'd been trying to explain it in words for years. The film explained it in eight minutes. She told me later that one donor increased his annual contribution by five thousand dollars after seeing it. That's impact you can measure in dollars.

Real estate virtual tours, event recaps, product demonstrations. I even made a film for a veterinary clinic about pet grief. They paid $900 and showed it in their waiting room on a loop. The vet told me three clients cried watching it and then booked appointments. That's ROI she could measure directly. Frank next door asked me why I kept making films about sad things. I told him people feel things when they're sad, and feelings make people act. He thought about that for a minute and said, "That's why those ASPCA commercials work." Frank's not as dumb as he looks.

Pricing this work is where most beginners fall apart. They charge by the hour because that's what feels fair. Don't do that. You're not a teenager mowing lawns. You deliver value, and value isn't measured in hours. When I started, I charged $50 an hour because that seemed reasonable for someone in New Jersey with a laptop. I worked ten hours on a project and billed $500. The client loved it. They showed it at a board meeting. Twenty people saw it. I made $500. The next client I quoted a project rate. $2,000 for the whole thing, deliverables included, two rounds of revisions. They didn't blink. I made four times as much for the same effort. Hourly billing punishes efficiency.

You get faster as you get better, which means you earn less the more skilled you become. That's backwards.

I use project rates now, with three clear tiers laid out on my services menu. The basic tier starts at $900 and covers a simple film — one location, one or two characters, under three minutes, stock music, basic color grade. Perfect for social media posts and simple explainers. The middle tier runs $1,800 to $3,200 and adds complexity — multiple scenes, custom music, more characters, some voice cloning, tighter editing. This is where most of my nonprofit and small business work lives. The premium tier starts at $4,000 and includes the full treatment: Digital Twins, cloned voices, multiple languages, original score, the kind of polish that plays well at board meetings and fundraising galas. I have a one-page document that I send to every prospect.

It has prices on it. Real numbers. Not "let's talk" or "it depends." Actual dollar amounts. When someone calls and asks what I charge, I email them the menu. Half the time they pick a tier immediately. The other half negotiate, which is fine — I built twenty percent padding into every price for exactly that reason.

The key insight about pricing AI work is what I call the AI efficiency premium. Because I can deliver faster than traditional production, some clients expect me to charge less. Wrong. I charge for the result, not the hours. If a traditional production company needs three weeks and fifteen thousand dollars to deliver something I can make in four days, the value to the client is identical. Actually, my speed is an advantage. They get their film faster. They can iterate faster. They can respond to market changes faster. Speed has value. Charge for it. You're not a discount version of traditional production. You're a different thing entirely — faster, more flexible, more iterative. That's worth a premium, not a discount.

Branded content is where the real money lives, and it's exploding right now. Every company on earth is trying to figure out short-form video for social media. They need dozens of pieces of content per month — Instagram Reels, TikToks, LinkedIn videos, YouTube Shorts. Traditional production can't keep up with that volume at a price that makes sense. A single traditional shoot day might cost five to ten thousand dollars when you factor in crew, equipment, location, catering, and post. That produces maybe three to five usable minutes of content. For that same five thousand dollars, I can produce twenty to thirty pieces of content across a whole month.

Different lengths, different platforms, different messages, all maintaining visual consistency because I'm using the same Digital Twin, the same voice, the same visual language.

I've been working with a local coffee roaster — eight videos a month for a flat retainer, each one featuring a consistent AI spokesperson (their Digital Twin, actually the owner's son who has a great face for camera), different scripts about their sourcing and roasting process, shot in virtual locations that look like Colombian farms and Italian piazzas. Total cost to me is maybe a hundred dollars in platform credits per month. They pay me fifteen hundred. That's fourteen hundred dollars in margin a traditional production company can only dream of. And because the turnaround is fast, if they have a new blend launching or a seasonal promotion, I can have a new video ready in forty-eight hours. Try getting that from a traditional crew.

Teaching and consulting is another revenue stream that flows naturally from having made a film. Here's the thing nobody tells you: if you've successfully made an AI film, you know more than ninety-nine percent of people who want to make AI films. The barrier to entry in knowledge is that high right now. Not because it's rocket science, but because it's new, fragmented, and changes monthly. Every tool updates its interface, its pricing, its capabilities. What worked six months ago might not work the same way today. People will pay you to compress six months of trial and error into a two-hour workshop.

I've done five workshops now at community colleges and arts centers — two hundred dollars per person, ten to fifteen people per session. That's two to three thousand dollars for a Saturday afternoon. I show them the actual workflow, the actual tools, the actual the sample short project files. I walk them through my HeyGen setup, my ElevenLabs voice cloning process, my Seedance prompts. They leave with a printed checklist and my email address. Several have become ongoing clients. One woman who took my workshop runs a small marketing agency in Jersey City. She hired me to consult on her first AI video project for a client. That single consultation led to three more referrals.

One-on-one coaching is the premium version of this. I charge five hundred dollars for a two-hour session where I review someone's project, troubleshoot their specific problems, and build them a custom workflow. Most of these come from people who saw the sample short and thought "I want to make something like that" but got stuck somewhere in the process. Usually it's the voice cloning — getting the AI voice to sound natural, not robotic. Sometimes it's making the Digital Twin not look like it came from the uncanny valley. Other times it's the editing —

they have great footage but can't make it flow. I enjoy the coaching. It's satisfying to watch someone have that breakthrough moment where it clicks.

My wife says I'm just showing off, but she says that about everything I do.

the second sample short became a portfolio piece in the most unexpected way. A funeral home in Hackensaw — that's not the real town name, but it's close enough — saw the film at a local arts screening and called me. The owner, a man named Gerald who wears suspenders and looks exactly like you'd expect a funeral director to look, said he wanted "something that shows we're human." Funeral homes have a marketing problem. Nobody wants to think about them until they have to. Gerald wanted a short film about the people who work there — the care they take, the small moments of connection. I made him a six-minute film using the same pipeline as the second sample short.

AI spokesperson, cloned voice, virtual locations that looked like his actual funeral home because I trained a model on photos he sent me. He paid $3,200. He plays it in his consultation room where families make arrangements. He told me three families specifically mentioned the film as something that made them trust him. That's a six-figure client relationship over time, all because I made a four-minute dark comedy about a whoopee cushion.

Your assets are reusable in ways that traditional filmmaking assets never were. Your Digital Twin isn't a one-time thing — it's a character you can cast in client work again and again, with different clothes, different settings, different purposes. Your voice clone works in any project. Your library of successful prompts — the ones that generated great backgrounds, good lighting, convincing crowds — that's intellectual property you built yourself. Your music library, your sound effects, your editing templates, your color grades. Every project makes the next one faster and cheaper. My fifth client project took half the time of my first because I wasn't starting from scratch. I was remixing assets I'd already built.

It's like Chuck's plumbing truck — he has every wrench, every fitting, every odd-shaped tool he might need, organized in drawers. When he gets to a job, he doesn't drive to the hardware store. He opens a drawer. My prompt library is my tool drawer.

Finding clients in the first place is the part nobody wants to talk about. You make a great film, you put it online, and then you wait. Nothing happens. I learned this the hard way. After the sample short, I posted it on Vimeo and assumed people would find it. They didn't. What actually worked was the unglamorous work of telling people it existed. I emailed the festival programmers I'd met. I posted in filmmaking groups. I reached out to local businesses and said, "I made this film using AI. Would you be interested in seeing how this technology could work for your organization?" Most said no. Some said yes. One yes became Denise. Denise became referrals. Referrals became a portfolio. Portfolio became credibility.

Pro Tip: Create that services menu document and update it quarterly. Add new services as you learn them, remove ones you don't enjoy. Always, always send a follow-up email within forty-eight hours of any festival screening, offering workshops. Most festival programmers will forward it to their mailing list. Build what I call a "film for nonprofit" practice — find causes you care about, offer them a discounted rate in exchange for portfolio rights and testimonials. Good cause, good portfolio piece, good karma. And invoice promptly. I send invoices the same day I deliver final files. Net fifteen payment terms. Most clients pay faster when they know you're organized. I use a simple invoicing app and accept credit cards.

Don't make it hard for people to pay you.

Common Mistake: Undercharging because "it only took me a weekend." That thinking will bankrupt you. The client isn't paying for your weekend. They're paying for the years of learning that let you deliver in a weekend. Another mistake: trying to be a generalist. Don't be the person who "does AI film." Be the person who "makes AI films for healthcare organizations" or "makes AI training content for manufacturers." Pick a niche. Own it. I started with nonprofits and grief counseling because the sample short established that credibility. Now I'm expanding into healthcare. But I didn't start everywhere. I started somewhere. The final mistake I see all the time: spending more time on your website than your work.

Your website is a business card, not a masterpiece. Make it clean, make it show your films, put your services menu on it, and then go make the next film. Nobody ever hired a filmmaker because their website had a really clever loading animation. My website cost me forty dollars and a weekend. It has my films, my services menu, and a contact form. That's it. It makes money every month.

The Possibilities Nobody's Talking About: What if your AI film studio became a subscription service? I'm serious. What if instead of selling individual projects, you offered "unlimited short films" for a monthly retainer? The Netflix model applied to content creation. Businesses need constant content. You need predictable income. A subscription solves both. Or what if you offered "film-as-a-service" — a flat monthly fee that covers concept, production, and delivery of whatever they need that month? I know a guy in Philadelphia doing this for real estate agents. Four videos a month, every month, like clockwork. He's making six figures. What if you became the person who trains other people's Digital Twins and then manages their ongoing content?

Every executive in America is going to want a Digital Twin within three years. Someone has to make them, maintain them, and troubleshoot them when the platform updates and the lips start looking weird. That someone could be you. That's not filmmaking in the traditional sense. That's something new. Something we don't have a name for yet. But the opportunity is real, and the window is open right now.

The truth is, I didn't get into this to run a business. I got into it to make films. But I discovered that making films and running a business aren't mutually exclusive. The business feeds the films. The films feed the business. Denise's call that day opened a door I didn't even know was there, and now I'm walking through rooms I never expected to enter. Next, I want to talk about what happens when you apply this same thinking to advertising — because branded content, it turns out, is where the economics get really crazy.

Chapter 19: Branded Content and the Generative Gold Mine

The first branded content project that came my way after the sample short wasn't from a big agency or a national brand. It was from a coffee shop. This little place in Montclair called Brewed Awakening — I know, the name is awful, but the coffee is genuinely great — and the owner, Marco, had seen the sample short because his girlfriend was in the audience at the festival screening. He called me and said, "I want something like that film you made, but for our Instagram." I said, "You want a twelve-minute thriller about grief and audio recordings for your coffee shop Instagram?" He said, "No, I want something that feels like that. Real. Human. Not an ad." I understood what he meant. the sample short had a quality that made people feel something.

Marco wanted that quality applied to his business. Not a sales pitch. A story that happened to feature coffee.

I made him a ninety-second film. A woman sits in his coffee shop early morning, headphones on, clearly going through something difficult. The barista — played by an AI avatar that looked enough like Marco to be recognizable — brings her a second cup without being asked. No words. Just a look between them. The tagline at the end: "Some things don't need an order." Marco cried when he watched it. Actually cried. He'd given me a few hundred dollars, some coffee shop photos to use as training data, and a rough idea. I gave him a film that made him feel proud of his business. That's the power of this pipeline. The same tools that made the sample short — HeyGen, ElevenLabs, Seedance, DaVinci Resolve — could serve both art and commerce.

The pipeline doesn't care whether you're making a festival drama or a coffee shop Instagram post. It just produces.

This is where I stumbled into something called Generative Creative Optimization, or GCO. It's a fancy acronym for a simple idea: using AI to create massive numbers of creative variations, testing them to find what works, and then scaling the winners. Traditional advertising works like traditional filmmaking. You plan one commercial, shoot it once, hope it works. If it doesn't, you spent ten to fifty thousand dollars learning that lesson. GCO changes the equation entirely. I can generate fifty ad variations in a day for what one traditional shoot costs in catering alone.

Here's the workflow. A brand comes to me with a brief — they want to promote their new product, reach a specific audience, achieve a specific outcome. I feed that brief into my AI pipeline and generate ten or fifteen different creative concepts in an afternoon. Different approaches. Emotional, humorous, informational, testimonial-style. The brand picks their favorites, and then we go to phase two: variation generation. Same core concept, but rendered with different faces, different backgrounds, different scripts, different music tracks, different color grades. Maybe version A features a young professional in a city apartment. Version B features a retiree in a suburban home. Version C features a parent in a minivan.

The underlying film structure is the same, but the packaging changes to speak to different audiences. Then we run them all as ads, measure performance, and double down on what works.

The the sample short pipeline translates beautifully to this work. the lead character — or a the lead character-like avatar — becomes a brand spokesperson. But instead of one the lead character in one setting, I can place him in twenty different environments. He's wearing a suit for the professional audience, a sweater for the cozy audience, scrubs for the healthcare audience. He's speaking to camera in English for the US market, Spanish for the Latino market, French for the European market. Each version takes minutes to generate, not days to reshoot. The economics are almost comical. A traditional brand video with this many variations would cost fifty thousand dollars minimum, probably more.

My cost to produce the full variation set is somewhere between five hundred and two thousand dollars, depending on how many languages and how much custom music. And because we can test everything, the client gets more data for their money. They know which version worked, not just that the campaign did "pretty well."

The tools line up like this. HeyGen provides the avatar spokesperson — consistent, recognizable, available 24/7. ElevenLabs handles voice cloning and localization — the same spokesperson sounds authentic in every language. Seedance generates the backgrounds and environments —

cityscapes, offices, homes, whatever the brief requires. And DaVinci Resolve ties it all together with color grading, music, and graphics. It's a branded-content assembly line that produces premium output.

But there's an ethical line here, and I'm going to talk about it because ignoring it will get you in trouble. Disclosure is not optional. Sponsored content must be labeled as such. AI-generated people in advertisements must be disclosed in many jurisdictions already, and that list is growing. The FTC is increasingly active on this. I always disclose. Always. My contracts include a clause requiring client disclosure, and I won't work with brands that push back on it. Not because I'm a saint — though my wife would disagree — but because it's not worth the risk. A fine from the FTC costs more than I'd make on ten projects. Plus, audiences aren't stupid. They can tell.

And when they catch you hiding that a spokesperson isn't real, the trust you've built evaporates instantly. Just disclose. It's cleaner. It's easier. And honestly, most audiences don't care that it's AI-generated as long as the content is good and you're honest about it.

Pro Tip: Start with one brand in one industry. Don't try to serve everyone on day one. I started with local food and beverage because Marco's project gave me a case study in that space. Then I used that case study to get a second restaurant, then a local brewery. Each project built proof for the next. Use your own films as proof of concept. the sample short demonstrated emotional storytelling. the second sample short demonstrated tone and humor. Both opened doors to different types of clients. Offer "ten variations for the price of one" as your differentiator. That's a number that gets attention in brand meetings. And always get performance data from your clients after campaigns run. Nothing sells the next project like hard numbers from the last one.

Common Mistake: Promising hyper-personalization you can't technically deliver yet. I've seen freelancers tell brands they can generate "a unique ad for every single customer." That's mostly not true at this point. You can do audience-segment personalization — different ads for different demographics. Individual-level personalization at scale requires infrastructure most independents don't have. Don't oversell. Another mistake: not understanding brand safety requirements. Major brands have strict rules about where their ads appear, what imagery is acceptable, what words can be used. Ask about this upfront. The final mistake: ignoring disclosure regulations because "everyone else is doing it." They're not.

And even if they were, that's not a defense that holds up in court or in the court of public opinion.

What If You Went Wild With This? The possibilities in branded AI content get genuinely science fiction if you let yourself imagine. What if every billboard showed a different AI-generated ad based on who was looking at it? Not just different content, but a different spokesperson — someone who matches the demographic of the person walking by. What if brands had AI-generated spokespeople who aged and evolved with the brand? A character who starts as a young parent in year one and grows into a grandparent by year ten, always consistent, always available, never demanding a raise. What if your studio became the "Netflix of branded content" — a subscription where clients pay a monthly fee for unlimited creative iterations?

They submit briefs, you generate variations, they test and optimize, and the cycle repeats. No traditional production bottlenecks. No waiting for crews to be available. No weather delays. Just creative concepts flowing directly into market-tested content. Marco's coffee shop was the beginning. But the model scales. A coffee shop today, a regional chain tomorrow, a national brand next year. The pipeline is the same. Only the invoice changes.

I think about that first call from Marco sometimes. A guy who wanted his coffee shop to feel more human, and he turned to AI to do it. There's poetry in that. The tool we thought would make everything robotic turned out to be the thing that let a small business owner tell a human story without spending a fortune. That's the thread that runs through all of this. The technology is just the tool. The humanity comes from what you choose to make with it.

Chapter 20: Selling Without Selling Your Soul

The worst pitch I ever made happened in the lobby of the Newark Airport Hilton, and I still think about it when I need to punish myself. I'd been emailing this woman, Linda, for three weeks — she ran programming for some regional cable network I'd found on LinkedIn at two in the morning while my wife was asleep and I was supposedly "just checking something quick." Linda had been polite enough to answer two of my emails, which in my desperate brain meant we were basically friends. She mentioned she'd be at a media conference in Newark. I decided to surprise her.

I want you to picture this. I'm wearing a blazer I bought for my nephew's bar mitzvah in 2017, slightly tight across the shoulders because pasta is non-negotiable in my life. I'm holding a thumb drive in a little velvet pouch I found in my wife's jewelry box like it's the Olympic torch. I spot Linda by the coffee station getting a cup of decaf.

I walked over and said — and I still can't believe these words came out of my mouth — "Linda? It's me. From the emails. The film guy. Jim. From New Jersey, obviously, because we're in New Jersey."

She looked at me the way you'd look at a man who just asked if he could borrow your toothbrush. Blank. Confused. Slightly alarmed.

I pressed on because I'm an idiot. "I made that twelve-minute thriller? With the sound engineer? You said 'thanks for sharing'?"

"Oh," she said. "Right."

She didn't remember. Of course she didn't remember. But I was committed now, drowning in my own terrible choices, so I opened the velvet pouch and presented her with the thumb drive like I was offering a royal gift. "I brought you the high-res version," I said. "In case you want to screen it. At your network. Or whatever."

She took the thumb drive, said something polite about being late for a panel, and walked away. I stood there by the decaf machine knowing I'd just done something that would haunt me forever. My wife still brings it up at dinner parties. She calls it "the velvet pouch incident." I deserve it.

But here's what I learned from that sweaty, humiliating morning. Direct outreach isn't the problem. I was the problem. I had no idea what I was doing, no idea what Linda actually needed, no idea how to talk to a buyer like a human being instead of a desperate salesman pushing velvet-pouch thumb drives on strangers in hotel lobbies. That encounter taught me more about selling films than any seminar ever could.

The Agent Myth

Let's get this out of the way first because it's the lie that keeps most independent filmmakers broke and waiting. You do not need an agent. Not for your short film. Not for your client work. Not for the twelve-minute thriller you made in your basement with HeyGen and a dream.

Agents don't care about you. I say this with love, having met a few agents at festivals over the years. Agents care about features with name actors. They care about packages they can sell to distributors who already have relationships with buyers. They're not interested in your AI-assisted short film that cost $847 and played at three festivals in New Jersey. That's not snobbery on their part — it's economics. They work on commission, usually ten to fifteen percent. Ten percent of zero is zero, and ten percent of the $500 you might make from a short film licensing deal barely buys them lunch in Manhattan.

Neighbor Frank once asked if he needed a real estate agent to sell his lawn mower on Facebook Marketplace. I told him no, just take good photos and write an honest description. You're selling a lawn mower, not a penthouse. Act accordingly.

This doesn't mean agents are useless — they have their place, and we'll get to that. But if you're sitting around thinking you can't sell your work because you don't have representation, you're making excuses. I sold the sample short to my first client while wearing a t-shirt with a coffee stain. No agent. Just me and the willingness to look stupid.

Finding Your Buyers

So where do these mysterious buyers live? How do you find the people who might actually pay money for what you've made? The good news is they're not hiding in a secret bunker. The bad news is they're busy, they're skeptical, and they get approached by dreamers all day long.

Film festivals are still the best networking opportunity most filmmakers ignore. Not the big ones — Sundance isn't calling about your AI short. I mean the regional festivals where you can talk to programmers over bad coffee in a church basement. At The sample short's first screening, I met a woman who ran corporate video for a medical device company. We talked for ten minutes about how I got emotional performances from AI-generated characters. Three weeks later she hired me for a training video. That happened because I was physically present, introduced myself like a normal person, and listened to what she needed instead of vomiting my life story at her.

LinkedIn is the other tool most filmmakers underuse. Not as a place to post about your "journey" — nobody cares about your journey. Use it as a research tool. Find decision-makers at companies that commission video content. Marketing directors. Communications managers. Content leads at agencies. Send them a connection request with a short, human message. Not "Dear Sir/Madam, I am a filmmaker seeking opportunities." Something like: "Hi Sarah, I saw your company's recent campaign — loved the editing. I make short films with AI tools. No pitch, just connecting." Then don't pitch them. Let them see your work and come to you. This takes patience, which most filmmakers don't have. That's why most filmmakers are broke.

Local business associations are the secret weapon nobody talks about. Your chamber of commerce, your rotary club, your industry meetups — these are full of people who need video content and have no idea how to get it made affordably. the second sample short was four minutes long and cost about thirty dollars to make. I showed it at a local business breakfast because someone dared me to. A funeral home director in the audience — I know, I know, but stay with me — approached me afterward. He said he'd been wanting to make a video that showed the human side of his business, something that wasn't depressing or sterile. the second sample short demonstrated that I could handle emotional material with a light touch. He hired me to make a short brand film for his company.

A funeral home. I made $2,400 on a project that took me four days. That's eighty times what the original film cost.

What Buyers Actually Want to Hear

Here's the thing that took me years to figure out. Buyers don't care about your rendering settings. They don't care that you used Seedance instead of Runway. They don't care about your "vision" or your "process" or how many hours you spent getting the lip-sync just right. They care about one thing: what problem do you solve for them?

When I pitched the sample short to that medical device company, I didn't start by talking about the lead character and his dead wife. I started by asking questions. What kind of training content were they making now? How much was it costing them? How long did it take? Turns out they were spending eighteen thousand dollars per training video using a production crew that took six weeks to schedule. I told them I could make something comparable in a week for under two thousand. Then I showed them the sample short not as a film, but as proof of concept. Look at the emotional performance in this scene. Look at the production quality. Now imagine this applied to your patient-care training module.

The shift is subtle but everything. You're not selling art. You're selling capability. The art is the demonstration, not the product.

Chuck the plumber came over to fix my garbage disposal last year and saw me working on a pitch deck. He asked what I was doing, and I explained I was trying to get someone to pay me to make a film. Chuck wiped his hands on his pants and said, "You ever see me try to sell a new boiler? I don't lead with the British thermal units and the copper alloy specs. I lead with 'your shower will be hot in the morning.' Same thing, Jim."

Chuck is a philosopher disguised as a plumber.

The One-Page Pitch Deck

Your pitch deck should be one page. One. If you can't explain why someone should care on a single sheet of paper, you don't understand your own film well enough to sell it.

This isn't laziness. This is respect for the buyer's time. They have eighty other things to do today. Your job is to make their decision easy, not to overwhelm them with your creativity.

My one-pager for the sample short had four elements. At the top, a strong still — the lead character at his mixing board, that navy sweater, the dim blue light on his face. Below that, a logline: "A grieving sound engineer discovers his dead wife's voice in unfinished recordings — and must decide whether to let her go." Then three short lines: twelve minutes, three festival screenings, seven hundred likes on the online premiere. At the bottom, my contact info and a QR code linking to the full film. That was it. No biography. No equipment list. Just the evidence and a way to reach me.

The response rate was about ten times better than the longer pitches I used to send. Some people scanned the QR code while I was standing there. That's the goal — get them to watch the work. The work sells itself if it's good enough.

the sample short as Pitch Tool

The most important sale I ever made with the sample short wasn't to a festival programmer or a YouTube audience. It was to a small marketing agency in Trenton that specialized in healthcare clients. The owner, a guy named David, had seen a clip I posted on social media. He emailed me cold — well, cold to him, warm to me because I nearly fell out of my chair when I saw an actual business inquiry in my inbox.

David didn't ask about my artistic process. He asked if I could make something similar for a hospital client that needed to humanize their palliative care program. Something warm instead of clinical. I told him I could. He asked my rate. I named $4,500, which at the time felt like saying "one million dollars." He said that was reasonable and asked for a proposal.

I sent the proposal that night because my wife drilled into me that you strike while the iron is hot. David approved it within forty-eight hours. I got a 50% deposit before I touched a single prompt. Delivered the eight-minute film in ten days. The hospital loved it, and David's agency hired me three more times that year. the sample short — an $847 film — led to over $50,000 in client work because it proved I could make emotionally resonant content quickly and affordably.

That's the pitch that matters. Not the one in the velvet pouch. The one where you show what you've done and explain what you can do next.

the second sample short as Calling Card

the second sample short was a different sales tool because it was so strange. A four-minute dark comedy about a funeral director inheriting pranks from his dead mentor? The whoopee cushion, the limerick eulogy, the coffin recording? On paper, nobody would pay for that.

But that's exactly why it worked. It showed range — the mix of melancholy and comedy that corporate video desperately needs but can never achieve because it's made by committee. The funeral home director who hired me said he knew within thirty seconds that I understood "the weird line our business walks every day." Your strangest work might be your best calling card. Don't hide the weird stuff. Lead with it.

Negotiation Basics

I'm from New Jersey, which means I'm direct to the point of rudeness, and I've had to learn to soften my edges without losing my spine. Here are the basics.

Name your price first. This is called anchoring, and it works. The first number mentioned becomes the reference point for everything that follows. Say "my rate is usually $3,000 but I'm flexible" and you've told them your rate is anything under $3,000. Say "the project will be $3,000" and you've anchored at $3,000. That second version gives you room to add value if they push back — "I could include a social media cut for $3,500" — instead of room to discount yourself into poverty.

Never work for exposure. Exposure doesn't pay rent. Exposure doesn't buy groceries. I've had people offer me "great visibility" in exchange for free work, and I used to say yes because I was flattered that anyone noticed me. Now I say no, politely but firmly. If they can't afford you, that's fine. Wish them well and move on. The only exception I make is for genuine nonprofits doing work I believe in, and even then, I trade my rate for something specific — a testimonial, a referral, a credit line that actually matters.

Get it in writing. Every single time. Even if the client is your cousin. Even if they're "definitely good for it." A simple email confirming the scope, the price, the timeline, and the payment terms counts as "in writing." I use a basic contract template I found online and modified over the years. Nothing fancy, just clear terms that protect both of us.

Always require a 50% deposit. Non-negotiable. I learned this the hard way when a client disappeared after delivery and I spent three months chasing $1,800 that I eventually gave up on because the stress was worth more than the money. Now I don't start work until the deposit clears. Most legitimate clients expect this. The ones who push back are telling you something important about how they do business.

When You Actually Do Need an Agent

I said you don't need an agent, and for short films and client work, you don't. But there comes a point where DIY stops making sense. If you make a feature-length film with real money involved — say, over $100,000 budget — an agent or sales rep becomes worth their commission. They have relationships you don't have. They know which distributors are actually paying right now. They can navigate international sales, which is a maze of territory rights and currency conversions and delivery requirements that will make you want to cry.

If you're negotiating with a streaming platform for an exclusive license, or navigating co-production agreements between companies in different countries, get someone who knows what those contracts should look like. I've never reached this level myself, but I know people who have, and the ones who succeeded knew when to stop doing everything themselves. Think of it like plumbing. I can fix a leaky faucet. I'm not touching the main sewer line. Know your limits.

Pro Tips

Create a one-sheet for every film you make. This is exactly what it sounds like — a single page with your poster image, logline, running time, technical specs, festival history if any, and your contact information. Keep it clean. Keep it professional. Hand it out like candy. I print mine at Staples on decent cardstock and carry them in a folder so they don't get bent. When someone asks about your work, you hand them a one-sheet. It shows you're serious.

Follow up within twenty-four hours of every meeting. Every single one. Send a brief email — "Great meeting you at the festival screening. Here's a link to the film we discussed. Let me know if you'd like to talk further." Most people never follow up, which means doing so immediately puts you ahead of 90% of your competition. My wife taught me this. She says the fortune is in the follow-up, which sounds like something on a motivational poster but happens to be true.

Join your local film commission's mailing list. Every state has one. They send out opportunities you'd never find — productions looking for crew, grants with deadlines, networking events. New Jersey's film commission has been surprisingly useful, and I say that as someone who assumes government organizations exist to send confusing paperwork.

Common Mistakes

I've seen smart filmmakers sabotage themselves in the same ways over and over. Here are the ones that make me want to shake people.

Apologizing for being AI-assisted. I hear this constantly — "Well, I made it with AI, so it's not really..." Stop. Stop right there. If the work is good, the work is good. How you made it is only interesting as a cost and speed advantage, not as a confession. I never open a conversation by explaining my workflow. I show the film. If they ask how I did it affordably, I explain. But I never apologize for being efficient. You don't apologize for using a power drill instead of a hand-crank, do you?

Undervaluing your work because you "didn't spend much." This one kills me. The value of your film has nothing to do with your budget. It has to do with what it does for the buyer. the sample short cost $847. The client who hired me because of it didn't care if it cost $8 or $8,000. They cared that it moved them, that it proved I could do something they needed. Price based on value, not your receipts.

Not asking for the sale. Filmmakers are terrible at this. We'll talk about our work for hours, we'll show clips, we'll explain our influences going back to film school, and we'll never actually say "would you like to hire me?" or "are you interested in licensing this?" The ask is terrifying but necessary. Practice it. Get comfortable with silence after you name your price. The first person who speaks after the number loses — another gem from my wife, who used to sell real estate and knows things about human nature that I don't.

Trying to befriend buyers instead of doing business with them. Be friendly, of course. Be pleasant to work with. But don't confuse a business relationship with friendship. Buyers can like you and still not hire you. They can hire you and not want to have dinner with you. The goal is professional respect, not best friendship. I made this mistake early on, spending hours trying to connect personally with people who just wanted to know if I could deliver on time and on budget.

What If You Went Wild With This?

Let me paint you a picture that sounds crazy until you think about it for more than thirty seconds. What if, instead of you chasing buyers, the buyers came to you?

Right now, traditional film production is slow and expensive. A studio needs a proof-of-concept to greenlight a feature? That used to mean spending fifty to a hundred thousand dollars on a short pilot. What if they came to you instead? What if your AI pipeline was so efficient that major producers hired you as a proof-of-concept service? You take their script and notes and deliver a polished scene in two weeks that looks like it cost a hundred times what they paid.

Or what if the agent model gets replaced by AI-powered matchmaking? Platforms that understand what buyers need and what filmmakers can deliver, connecting them directly without the ten percent haircut. The technology is clumsy now, like a baby deer learning to walk. Give it five years.

And here's what keeps me up at night in a good way. What if your efficiency becomes your brand? Traditional production companies need a crew of twenty and a month of prep to deliver what you can make in your home office in a week. That's an unfair advantage. The market always rewards speed and quality at lower cost. The plumber who showed up same-day when my basement flooded got my business forever. The one who said "we can come Tuesday" got nothing.

The velvet pouch days are behind me. I hope. But the lesson remains — you don't need permission to sell your work. You don't need an agent's blessing. You need good work, clear communication, and the willingness to hear no a lot of times before you hear yes.

Before you sell anything, though, you need to make sure you actually own what you're selling. There's a whole new frontier of legal questions that come with AI filmmaking, and ignoring them won't make them go away. Trust me, I learned that the hard way.

Chapter 21: The Legal Stuff Without the Panic

The handshake deal that almost ruined me happened in the parking lot of a Chili's in Paramus. A guy named Gary had seen the sample short at a local screening and wanted to "distribute it." That's the word he used. Distribute. I heard that word and something primal lit up in my brain. Like Miramax was calling.

We met at Chili's because Gary said he "did his best thinking over a Tex-Mex salad." That should have been my first warning. His plan was to put the sample short on his "platform" — a website he ran from his basement that got about forty visitors a month. In exchange, he'd "handle everything" and we'd "split the profits." I shook his hand and thought we'd made a deal.

Two weeks later, Gary had uploaded my film to three platforms I'd never heard of, slapped his logo on the opening credits, and was running pre-roll ads for products I didn't want associated with a film about grief. When I called him to ask what the hell he was doing, he said, "It's in the contract, Jim."

"What contract? We shook hands in a Chili's parking lot."

"That was our agreement. A handshake is a contract."

It took three months and a $600 consultation with an entertainment attorney to untangle myself. I got lucky because Gary was sloppy. If he'd been smarter, I might have lost control of work I'd spent months creating.

The lesson wasn't just "get it in writing" — though obviously, get it in writing. The lesson was that legal protection isn't something you worry about when you get successful. It's something you worry about now, because the world is full of Garys, and AI filmmaking adds a whole new layer of complexity.

I need to be very clear before we go any further: I'm a filmmaker, not a lawyer. Everything in this chapter comes from my own experience, my own research, and my own expensive mistakes. Laws change. Platform terms of service change monthly, sometimes weekly. What's allowed today might be restricted tomorrow. Before you make any commercial decisions based on what you read here, consult a qualified entertainment attorney. Think of that consultation fee as insurance. Insurance against much bigger problems that could cost you your work, your money, or your peace of mind.

What You Actually Own

Let's start with the good news. You own more than you think.

Your script is yours the moment you write it. Copyright law protects original works from the instant they're fixed in a tangible medium — typed into Google Docs, scribbled on a napkin. Registration with the Copyright Office just gives you stronger protections if someone steals it.

Your edit is yours because your creative choices are uniquely human. The decision to cut from the lead character's face to the oscillating waveform at exactly that moment — no AI made that. You did. The Copyright Office cares about those choices. They're what make your work yours even when some elements were generated by machines.

Your prompts are yours too. The specific text you typed into HeyGen or Seedance — that's your creative direction. It's like being a photographer who sets up the shot, chooses the lighting, frames the composition, then presses the button. The camera takes the picture, but the creative decisions are yours. Document your prompts. Save them. They're evidence of your contribution.

Your sound design, your music choices, your mixing decisions — all yours. I spent fourteen hours getting the audio layering right in the sample short, building the absent wife's voice into the mix so it felt present without overwhelming the scene. The tools helped, but the judgment was mine.

What You Don't Own

Now the trickier part. There are things in your AI-assisted film that exist in a gray area, and you need to understand the boundaries.

AI-generated faces don't belong to anyone. That's both good and bad. It's good because nobody can sue you for using a generated face that looks like them — because it isn't them. The face was created from statistical patterns across millions of training images, and it doesn't depict any real person. It's bad because you also can't claim ownership of that specific face. It's not your intellectual property. It's a tool output, like using stock footage. You own the film you put it in, but you don't own the face itself.

AI-generated voices are governed by the platform's terms of service. When I created the lead character's voice in ElevenLabs, I agreed to their terms — commercial use allowed, with restrictions. But those terms could change. Probably will change. Platforms are figuring out their business models in real time. Read the terms. Screenshot them. Keep them organized. I have a folder called "Legal — Platform Terms" with dated screenshots. My wife thinks I'm paranoid. I think I'm prepared.

AI-generated environments from platforms like Seedance or Runway fall under the same logic. You own your creative composition — the prompts you wrote, the sequences you arranged, the edits you made. But the raw generated elements? Those are outputs licensed to you under the platform's terms, not creations you own outright. It's a distinction that matters more as your work becomes more valuable.

The Consent Framework

This is the part where I get serious, because consent isn't just legal protection — it's moral protection, and the two are deeply connected.

Every voice clone needs documented permission. If you clone a real person's voice — an actor playing a role, a family member providing narration, yourself even — get it in writing. Describe exactly what you're going to use the voice clone for, how long the project will be distributed, and what platforms it will appear on. Have them sign it. Keep the signed document forever.

HeyGen's Digital Twin consent process is actually the gold standard here, and I don't say that lightly because I complain about most platform processes. When you create a Digital Twin of a real person in HeyGen, the person has to go through a consent flow on camera. They verify their identity. They explicitly agree to the creation of their digital likeness. This creates an audit trail that protects both you and the person being cloned. If someone ever claims you used their likeness without permission, you have documented proof that they consented through a verified process.

For the sample short, the voice actor who played the lead character came to my basement studio and recorded about twenty minutes of dialogue. Before we started, I handed him a release form my attorney had reviewed. It specified that I'd use his recordings to create an AI-generated voice clone for the film, that the film would be distributed online and at festivals, and that he granted me these rights in exchange for our agreed session fee. He read it, signed it, and we made great art together.

a supporting character's actress — the woman in the green cardigan — signed a similar release covering her visual likeness in the Digital Twin process. Same with the child character. I have written consent from every human performer whose likeness or voice appears in that film, even indirectly through AI generation.

The rule is simple: when in doubt, get it in writing. When not in doubt, get it in writing anyway. Chuck the plumber has a sign in his truck that says "verbal agreements lead to verbal disputes." He's right about that too.

Music Rights

Music is where a lot of AI filmmakers get themselves into trouble, and the situation is genuinely confusing because the rules are evolving faster than anyone can track.

AI-generated music from platforms like Suno and Udio currently has uncertain commercial status. These services are fighting legal battles right now about whether their training data was properly licensed. The terms of service might say one thing while a court ruling next year says something completely different. My personal approach — and this is just my approach, not legal advice — is to avoid using AI-generated music for commercial projects where the music will be prominently featured.

Instead, I license music from Artlist and Epidemic Sound. These are subscription services that give you clear commercial licenses for a predictable annual fee. I know exactly what I'm allowed to do with their music. I can use it in client work, in festival screenings, in online distribution. There's no ambiguity. I pay my subscription every year and I sleep well at night.

Grumpy Old Man Tool Note: Artlist and Epidemic Sound

Artlist and Epidemic Sound are music licensing services, and they exist because the phrase 'I found this song online' is not a legal strategy. Music rights can ruin your day fast. A clear license is not exciting, but neither is a cease-and-desist letter arriving after your film finally gets attention.

Use licensed music when the scene needs reliability. AI music can help explore tone, but licensed tracks can provide safer commercial use, better polish, and predictable rights when the project is going public, entering festivals, or serving a client. The smart workflow is simple: temp with AI, test emotion, then decide whether to keep, replace, license, or commission.

The limitation is laziness. Do not pick the first cinematic piano track because it made you feel deep for eleven seconds. Music should serve the scene, not pour syrup over it. Also read the license. I know, reading terms is boring. So is bankruptcy court. Choose your boredom.

For the sample short, I used licensed ambient music from Artlist and created atmospheric sound design from royalty-free elements and original recordings. The ticking clock in the lead character's studio? That's my actual kitchen clock, recorded on my phone. I own that sound completely. Sometimes the best solution is to grab a microphone and make your own audio.

If you use AI-generated music, read the current terms carefully. Screenshot them. If any terms about commercial use or redistribution are unclear, don't use it for commercial work. The money you save isn't worth the legal headache.

The Human Authorship Requirement

This is the big one, and it's going to become more important every year. The US Copyright Office has made clear that purely AI-generated works are not eligible for copyright registration. There needs to be sufficient human creative input for a work to qualify for protection.

What counts as "sufficient"? That's the fight everyone's having right now. What we know is that your creative choices matter — your script, your shot selection, your editing, your direction of AI outputs, your sound design. These human elements, taken together, can make your AI-assisted film eligible for copyright registration.

The key is documentation. Save your script drafts with timestamps. Screenshot your prompts with dates. Keep project files organized. If you ever need to defend your copyright, you'll need to show exactly what you contributed versus what the AI generated.

I registered the sample short with the US Copyright Office for $65. I was honest on the form — described my role as writer, director, and editor, noted the AI-generated visual elements created under my direction, and specified that the script, editing, sound design, and music were my original human work. They registered it. That $65 gives me a public record of ownership and the ability to sue for statutory damages. Worth every penny.

Terms of Service Compliance

Every AI platform has its own rules, and they are not the same. What HeyGen allows is different from what ElevenLabs allows, which is different from what Seedance allows, which will probably change next quarter because this entire industry is moving at the speed of a toddler on a sugar high.

You need to read the terms. I know, I know. They're boring. They're long. They're written by lawyers for lawyers. But buried in those terms are the actual rules that govern what you can and cannot do with your own work. Can you use the output commercially? Can you sell it to a client? Can you enter it in festivals? Can you distribute it on streaming platforms? The answers are in there, and "I didn't read the terms" is not a defense if you get sued.

I have a system now. Every time I start using a new platform, I screenshot the terms of service page and save it with the date. When terms change, I compare new versions to old ones. This sounds obsessive because it is. But obsessing about this stuff is cheaper than obsessing about a lawsuit.

Seedance allows commercial use of generated video but prohibits using outputs to train competing models. HeyGen's terms cover Digital Twin creation with specific consent requirements. ElevenLabs has tiered permissions — Professional Voice Cloning has different rules than standard voices. None of this is insurmountable. It's just paperwork. You wouldn't skip recording audio because it's tedious, right? Don't skip reading the terms because it's boring.

The sample short's Legal Preparation

Let me walk you through exactly what I did for the sample short, because concrete examples beat abstract advice every time.

The voice actor who played the lead character signed a work-for-hire agreement and a voice clone release. The work-for-hire established that I commissioned his performance and owned the recordings. The clone release granted me permission to create an AI-generated voice through ElevenLabs for the film. Both documents were reviewed by my attorney. That $340 legal review is part of the $847 total budget for the sample short — not an extra expense, just part of making the film properly.

a supporting character's actress signed HeyGen's Digital Twin consent plus my broader performer release covering distribution. the child character's mother signed parental consent since she's a minor. the absent wife, the dead wife, was played by a different actress who recorded the audio fragments; she signed the same releases. The Photo Avatar of the absent wife in the framed picture on the lead character's shelf was created with HeyGen's face generation, so no real person's likeness was involved. That's noted in my legal folder.

I also have documented records of every AI-generated element. Which platform generated which scene. Which prompts I used. Which terms version was current. It's all in a folder called "the sample short — Legal" backed up in three places because I'm not losing years of work to a hard drive failure.

the second sample short's Different Legal Profile

the second sample short was completely different — I used AI-generated faces for every visible character. No human actors to release, no Digital Twins to consent. Harold Vance, the funeral director with the progressively disheveled black suit, was fully AI-generated. His face belongs to no one. His voice came from an ElevenLabs platform voice with pitch adjustments I made.

This meant no performer releases, but total dependence on platform terms. Every visual came from HeyGen or Seedance. Every audio element came from ElevenLabs or my sound design. I needed to verify that all three platforms allowed commercial use, festival submission, and online distribution.

The advantage is speed — I made the second sample short in three days. The disadvantage is building on a foundation you don't fully control. If platforms change terms retroactively, you could be in a difficult position. My protection strategy was the same: screenshot the terms, document everything, keep records organized. The film cost about $30 to make, but I spent ninety minutes reading terms before I started. That was the most important time I spent on it.

Pro Tips

Register your copyright with the US Copyright Office. It costs $65 and gives you a public record of ownership that predates any dispute. If someone steals your work, registration lets you claim statutory damages — you don't have to prove exactly how much you lost. The process is online now, and while the interface looks like it was designed in 1998, it works. I've registered five films this way, and each time the certificate arrives, I feel like a grown-up filmmaker.

Keep a legal folder for every project with all signed releases, platform terms screenshots with dates, contracts, and correspondence about rights. Organize it the same way every time. My structure: Project Name, then subfolders for Releases, Platform Terms, Contracts, and Copyright Registration. If someone asks me to prove I have rights to something, I can produce documentation in under five minutes. Professionalism is often just being organized.

Consult an entertainment attorney before any deal over $5,000. That's my personal threshold for "this is getting serious enough." The attorney who untangled the Gary situation now reviews my client contracts, and I build her fee into project quotes. Clients don't mind paying for legal review when you explain it protects both parties.

Watermark your review copies. When you send a film to anyone who hasn't paid you, put a visible but non-intrusive watermark on it — your name or email, semi-transparent. It won't stop a determined thief, but it stops casual theft. I learned this after someone forwarded an unwatermarked review copy of my work to a third party who claimed they'd never heard of me.

Common Mistakes

Assuming AI-generated means "no legal issues" is the most dangerous mistake I see. Because no human actor was photographed, people think there are no legal questions. Wrong. The questions are different, not absent. You still need to understand platform terms, document your creative contributions, and protect your work from infringement. AI doesn't erase legal complexity — it just changes the shape.

Not reading platform terms because they change is lazy. "They change anyway, so why bother?" You bother because you need to know today's rules. The terms are a contract. Would you sign a contract without reading it? Don't answer that — I know some of you would. Stop doing that.

Using someone else's voice without written permission can destroy your project and reputation. I'm not talking about platform-generated voices — I'm talking about cloning your friend or relative. If their voice is identifiable and you don't have explicit written permission, you're exposing yourself to claims of misappropriation of likeness and fraud. Get the release. The conversation might be awkward for five minutes. The lawsuit will be awkward for years.

Using music without checking commercial rights has derailed more indie films than bad reviews. That free track? That Creative Commons license? Each has specific constraints. Read the license. When I need music I can use anywhere, I pay for Artlist. The certainty is worth more than the money saved on free alternatives with ambiguous terms.

Signing contracts without understanding what you're giving away is how you end up like me with Gary. Every contract has a rights grant section — where you give the other party permission to do things with your work. Are you giving exclusive rights or non-exclusive? For how long? In what territories? Can they alter your work? These are the heart of the deal. If you don't understand the language, hire someone who does. That $400 for a contract review could save you from giving away your film forever.

What If This Changes Everything?

The legal terrain around AI filmmaking is going to transform this decade. Here's what I think about when I'm lying awake at three in the morning, which happens more now that I'm sixty-five and my prostate has opinions about sleep.

What if AI filmmaking forces a total rewrite of copyright law? We're applying twentieth-century frameworks to twenty-first-century tools, and the mismatch gets worse monthly. What level of human involvement qualifies a work for copyright when AI tools are involved? Do AI companies owe compensation to creators whose work trained their models? These questions will get answered messily, and the answers will reshape everything.

Then there's the Digital Twin question. HeyGen's consent-based model is just the beginning. What if your Digital Twin outlives you? What if I create a verified digital version of myself and explicitly grant my heirs the right to license that Twin after my death? Could my grandchildren earn residuals from my digital performances decades from now? That sounds like science fiction until you realize that estates of dead actors already license their likenesses for CGI appearances. Posthumous digital performance rights are going to be a massive legal specialty, emerging from the tools we're using today.

And here's the one that really gets me. What if "synthetic performer rights" becomes a whole new legal specialty? As AI-generated characters become more common and more valuable, someone's going to have to figure out how to protect them. If I create an AI character that becomes popular — say, a recurring digital actor in a series of short films — do I have trademark protection? Can someone else generate a similar-looking character and claim it's different because the underlying pixels don't match? We're going to need legal frameworks for synthetic identity that don't exist yet, and the lawyers who figure this out first are going to make a fortune.

The legal side of AI filmmaking isn't just about avoiding trouble. It's about seeing around corners. The filmmakers who take this seriously now — who document everything, who read the terms, who get proper releases, who register their copyrights — are going to be positioned perfectly when the legal terrain matures. Everyone else is going to spend the next ten years scrambling to catch up, or worse, getting sued out of existence because they assumed "AI" meant "no rules."

Gary from the Chili's parking lot taught me one good lesson, even if he taught it badly. Your work has value, and people will try to take it if you don't protect it. The tools have changed. The stakes haven't. Protect your work. Document your process. Read the terms. Get it in writing. Spend the money on legal advice when you need it. Think of it like insurance — boring until you need it, and then it's the only thing standing between you and disaster.

My attorney's name is Rachel, by the way. She's thirty-four, she wears sneakers to client meetings, and she charges me $350 an hour to review contracts. She's worth every penny, and I'm not just saying that because she got me away from Gary. I'm saying it because sleeping well is worth paying for.

Chapter 22: OpenClaw, Hermes, and the Rebel's Toolkit

Grumpy Old Man Tool Note: OpenClaw and Hermes

OpenClaw and Hermes represent the rebel's toolkit - the kind of systems people look at when they want more control, more automation, more local workflow, or more advanced command over AI production. This is not where every beginner should start. A beginner needs a finished scene, not a science fair volcano made of APIs.

The possibility is serious production infrastructure. Imagine one system tracking scripts, prompts, assets, folders, model outputs, revisions, captions, marketing copy, and file delivery. Imagine a command center that helps move material from writing to visuals to voice to edit to release. That is the future: not one magic button, but connected tools behaving like a small production office that does not need donuts.

The limitation is complexity. Advanced systems can waste your time if you build them before you know your workflow. First make something. Then automate what repeats. Do not build a spaceship to cross the street. But once you know the route, tools like OpenClaw and Hermes point toward the kind of command-center filmmaking that small creators should be paying attention to.

The first time I tried to run OpenClaw on my laptop, it felt like asking my toaster to perform open-heart surgery. I'd spent three hours reading the GitHub readme — which was written by people who clearly think "user-friendly" means "only requires you to edit three configuration files in Python." I finally got the command typed in, hit enter, and watched my screen fill with red text faster than a blood pressure monitor at tax time.

I was up until two in the morning. At 2:17 AM, I posted a screenshot of the error to a Discord server filled with people whose usernames looked like someone fell asleep on a keyboard. "Help," I wrote. "65-year-old filmmaker. Just want to make a talking head. Getting this nightmare instead."

At 6:42 AM, a reply came in from klaus_dev in Hamburg. "Your PyTorch version is mismatched with your CUDA drivers. Downgrade to 2.0.1. Also, your GPU is too old for the default model — use the lightweight flag." I followed his instructions. Twenty minutes later, I had a generated face staring back at me. The lip sync looked like the person was chewing a difficult sandwich. But it worked. And it was free.

That's open-source AI filmmaking in a nutshell. It will humble you. Then it will hand you tools that would have cost a fortune six months ago.

What OpenClaw Actually Is, and What It Isn't

OpenClaw is the closest thing the open-source world has to a full avatar and voice generation pipeline. Think of it as the community's attempt to build what HeyGen and ElevenLabs sell — only without the subscription fee and with a whole lot more swear words in the documentation.

It covers text-to-speech, voice cloning, lip-sync animation, and some video generation. None of it works as smoothly as the paid tools. But the core technology is surprisingly capable, and it gets better every month.

The voice cloning in OpenClaw is the strongest piece. It's not as crisp as ElevenLabs PVC — you won't get the emotional nuance that the lead character's voice needed for the sample short. But for narration, for background voices, for characters where perfection isn't the point, it's genuinely useful. I've used OpenClaw-generated voices for radio commercials and temp tracks.

The avatar generation is where it falls down hardest. OpenClaw has the rendering quality of a video game cutscene from 2019. If you need a client-facing avatar that represents a brand, it's not your tool. But if you're making a sci-fi short where the protagonist is an android trying to pass as human? That slight uncanny quality becomes a feature, not a bug.

Hermes and the Hacker Ethos

Hermes is newer and wilder. It's a text-to-video and image-to-video pipeline adopted by a Discord community that seems to run on energy drinks and righteous fury. The video quality is uneven. Sometimes you get a shot that looks cinematic and haunting. Sometimes you get a figure with three arms walking through a house where the laws of physics have taken a holiday. But it runs on your local machine. No per-minute fees. You can generate a hundred variations of a shot and the only cost is your electricity bill.

The community aspect is what makes it special. Someone posted a complete workflow for generating atmospheric B-roll on Reddit. Another user built a script that filters out generations with obvious deformities. These aren't employees of a company. They're people who like solving problems and sharing solutions.

The Trade-Off Nobody Wants to Talk About

Here's the hard truth about open-source filmmaking tools: they cost time instead of money. When HeyGen goes down, I email support. When OpenClaw breaks, I'm reading error logs at midnight like I'm studying for an exam I never signed up for.

You need to be comfortable with the command line. Not an expert — I'm certainly not — but comfortable enough to type commands without hyperventilating. Chuck the plumber watched me google error codes for forty-five minutes one day. "This is like me fixing a boiler by reading German engineering diagrams," he said. "Possible. Not how I'd choose to spend a Saturday." But Chuck got it too. "When you figure this out, you don't owe anybody a monthly payment. That boiler company charges me three hundred bucks for diagnostic software. Yours is free."

The One-Person Production Company

Open-source tools make the one-person production company financially possible. I run NJ Film Productions out of my house. No office rent. No payroll. No investors. Just me, my laptop, paid subscriptions for client work, and open-source tools for personal experiments.

My rule is simple. If a stranger is paying, I use tools that guarantee professional output. If I'm the only one watching, I use whatever teaches me something. the sample short needed HeyGen's polish. the second sample short was a thirty-dollar experiment where rougher open-source B-roll of a funeral home worked perfectly — the slight weirdness matched the dark comedy.

The open-source community moves faster than any company. A technique requiring a paid subscription today might be free in six months. But paid tools offer reliability open-source can't match. You need both.

Finding Your People

The OpenClaw Discord server has a channel called "dumb-questions" and it's the most honest place on the internet. People post broken code and confused screenshots, and other people answer. Not always politely — there's a guy named sysop_mike who responds to half the posts with "read the docs first" — but help comes.

GitHub is where the tools live. Reddit's r/LocalLLaMA and r/MediaSynthesis are where people share what they've actually made. Not marketing demos. Real work, failures and all. The transparency is refreshing after the polished gloss of commercial websites.

Pro Tip: Start with one open-source tool. Not five. One. I started with OpenClaw's text-to-speech module because it was the simplest piece. Once that worked, I got curious about what else it could do. Curiosity is a better teacher than ambition. Trying to replace your entire paid stack in a weekend is how you end up crying into a keyboard at two in the morning with seventeen browser tabs open and nothing working.

Pro Tip: Keep your paid subscriptions active while you learn the open-source alternatives. Don't cancel HeyGen because you're "going to figure out OpenClaw." I tried that. Then a client called with a rush job, I couldn't get OpenClaw working in time, and I looked like an amateur. Use open-source for personal projects first. Prove it works for you before you bet a deadline on it.

Common Mistake: Not reading the documentation. I know. The docs are badly organized, half-finished, and assume you already know things you don't. Read them anyway. Most of the "breaking" I've done with open-source tools came from skipping a setup step that was clearly written somewhere I didn't bother to look.

Common Mistake: Updating without backing up. Open-source tools change constantly. Someone pushes a new version, you install it excitedly, and suddenly nothing works anymore. Before any update, copy your entire working folder somewhere else. Name it "openclaw_working_september_DONT_TOUCH." I have about twelve of these folders now. At least three of them saved me from total disaster.

What If You Went Wild With This?

Here's what keeps me up at night thinking — in the good way, not the error-message way. What if someone built a complete filmmaking pipeline using only open-source tools and gave it away? Not sold it. Gave it. A downloadable package that lets anyone with a decent laptop write, generate, edit, and export a short film without spending a dime on software. We're closer to that than most people realize. The pieces exist. Someone just needs to connect them with documentation that normal humans can read.

What if the next great filmmaking tool isn't built by a corporation at all? What if it emerges from a Discord server at three in the morning, coded by someone in Estonia who just wanted to solve a problem they had while making their own film? That's how the best open-source tools have always worked. Someone scratches their own itch, shares the solution, and it turns out thousands of other people had the same itch.

And here's the one that really gets me. What if you made a feature film for the cost of electricity? Not a gimmick. Not a proof-of-concept. A real feature — ninety minutes, story and characters and music — that you made in your house, on your laptop, using tools that are free. That would have sounded like science fiction when I started making films. Now it sounds like a project I could start next month if I was a little braver and a little more comfortable with Python.

The stranger from Germany never asked me for anything. Not a subscription, not a thank-you, not even my name beyond "65-year-old filmmaker." He just saw someone confused, knew the answer, and shared it. That's the open-source ethos. And it's why, despite every frustrating error message, I keep coming back.

Chapter 23: The Stigma — and Why It Doesn't Matter

The worst thing anyone ever said to me about an AI film happened at a Q&A in a church basement in Montclair. I'd screened the sample short at this tiny festival — fourteen people in folding chairs, a projector that buzzed loud enough to compete with the soundtrack, the whole charming disaster of independent film. After the lights came up, a guy in the second row raised his hand. He looked to be about my age, maybe a little younger, wearing a jacket with leather patches that made me think he taught film studies at a community college somewhere.

"I have a question," he said. "When are we going to stop pretending this counts as filmmaking?"

The room got very quiet. Not the good kind of quiet you get after an emotional ending. The uncomfortable kind.

"You used computers to generate your actors," he continued. "You didn't light anything. You didn't direct a performance in any meaningful sense. You wrote some prompts and pressed buttons. I'm genuinely asking — what separates what you just showed from a video game cutscene? And why should anyone take it seriously as cinema?"

I stood there with a microphone in my hand feeling like I'd been punched in the stomach. Fourteen people watching me get dismantled by a stranger in a church basement. I mumbled something — I honestly don't remember what — and the moderator moved on to the next question. But I thought about that man's words for weeks afterward. I woke up at three in the morning replaying them. I told my wife about it and she said, "Well, did the film move people or not?" That was the right question. But it took me time to see it.

The Stigma Is Real, and It Will Find You

If you make films with AI tools, people will tell you — to your face, online, at family dinners, in Q&As — that what you're doing isn't legitimate filmmaking. They'll call it cheating. They'll say it's soulless. They'll invoke the ghost of actual artists who suffered for their craft, implying that because you didn't spend six hours in freezing rain waiting for the light to change, you haven't earned the right to tell stories.

Neighbor Frank, who talks too much but means well, cornered me in my driveway last spring. He'd read something online about AI filmmaking. "Jim, I gotta be honest with you," he said, which is never the start of a conversation you want to have. "It feels like you're playing with toys instead of doing real work. My nephew's a cinematographer. He went to school for it. What you're doing — isn't that going to put guys like him out of business?"

I wanted to tell Frank to mind his own business, which is my default response to Frank about most things. But he was asking something real. He wasn't being cruel. He was confused, and a little scared on his nephew's behalf, and trying to understand why someone he'd known for twenty years was suddenly making films without a camera crew.

That conversation stuck with me because it captured something I couldn't shake. Frank wasn't a troll. He was a neighbor who'd watched me build a deck, fix a fence, and make a low-budget film the traditional way. His confusion wasn't hostility — it was the disorientation of watching someone you trust adopt tools you don't understand. And that meant I needed a real answer, not a defensive one.

Where the Criticism Comes From

I've spent a lot of time thinking about why AI filmmaking provokes such strong reactions, and I've sorted the critics into four rough categories in my head. Understanding which one you're dealing with helps you decide how to respond.

Some people are afraid. They're crew members, technicians, union folks who've spent decades building skills that paid their mortgages. To them, AI feels like a threat to their livelihood, not a new set of tools. These people deserve empathy, not arguments. I was never going to hire a full camera crew for the sample short — I made it for eight hundred and forty-seven dollars. But that doesn't mean the fear isn't real for people who work on bigger productions. The camera operator who worked his way up from PA to union member over fifteen years isn't wrong to worry. His life is built on a set of assumptions that technology is challenging. Telling him not to worry doesn't help.

Acknowledging that his concern is valid, even while you pursue your own path, is the only honest response.

Some people are purists. They believe there's a right way to make films and it involves celluloid, or at least a camera with a lens that costs more than a car. These people also believe vinyl sounds better, which it doesn't, but arguing with them is pointless. Their position is aesthetic, not practical. You can no more convince them that AI-generated performances are valid than you could convince my grandmother that microwaved food was edible. She died still believing that.

Some people genuinely don't understand what the process involves. They think you typed "make a sad movie about a sound engineer" into a chat box and eleven minutes later uploaded a finished film. They have no concept of the scripting, the voice direction, the sound design, the editing, the color work, the music, the fifty decisions per minute that go into every frame. Educating these people is possible, if they're willing to listen. But you can't educate someone who thinks they already know everything.

And some people raise legitimate ethical concerns. Consent for training data. Disclosure to audiences. Labor practices at the AI companies themselves. These concerns are real, and if you're going to work in this space, you need to think about them honestly. Brushing them off doesn't make you a bold innovator. It makes you lazy. I've thought about the training data question a lot. Models learn from something. Knowing what that something is, and whether the people who created it consented to its use, matters. It doesn't mean you can't use these tools. It means you should know what you're participating in and make your own peace with it.

How I Respond Now

After months of stewing about the man in the church basement, I figured out my answer. Not the answer — there's no such thing — but my answer. I lead with the story, not the tool.

the sample short isn't about HeyGen. It's about the lead character losing his wife and finding her voice in static. The grief is what matters. The audio recordings as metaphors for memory — that's what matters. Whether the lead character's face was rendered by an algorithm or captured by an Arri Alexa doesn't change what the story is about. When Spielberg used a Steadicam in Raiders of the Lost Ark, nobody asked whether the movement was legitimate. They asked whether it served the story. Technology is a choice, not a confession.

When people ask me now how much of my film is AI, I deflect. Not because I'm hiding anything — I disclose when it's appropriate — but because the question is usually asked from a framework that doesn't interest me. "I used every tool that served the story," I say. "The same way I've always used every tool that served the story. The only difference is these tools are new, and people are still deciding how they feel about them."

Sometimes that satisfies them. Sometimes it doesn't. I've learned to live with the discomfort of not convincing everyone. Not everyone needs to agree with you for your work to be valid. That's a hard lesson for people who grew up wanting approval — which is most of us, if we're honest.

Disclosure: When to Tell and When It's None of Their Business

This matters, so listen closely. You should disclose AI use when you're asked directly — festival submission forms are starting to include this question, and lying on a submission is a good way to get blacklisted from that festival forever. You should disclose when a client has a right to know what's in their deliverable. If someone's paying you to make a training video and they specifically want to know how it was produced, tell them. It's their money.

You should disclose in educational contexts, where people are learning and have a right to understand the full picture of how something was made. If you're teaching a workshop and showing your film, be honest about your process. Students deserve transparency. I showed the sample short to a high school media class last fall and I walked them through the whole pipeline — HeyGen for the characters, ElevenLabs for the voice, Seedance for the B-roll, DaVinci for the edit. A girl in the front row asked if she could do the same thing on her Chromebook. I told her some of it, yes, and some of it she'd need access to tools her school might not have yet. That was the truth. That's all disclosure requires.

But your creative process is yours. No painter lists every brush they used. No writer credits their word processor. If someone watches your film cold and responds to it emotionally, you don't owe them a technical breakdown of your pipeline. Disclosure becomes self-sabotage when you lead with it — when you apologize before anyone has even seen the work, when you warn audiences that they might not think it counts. Have some confidence. Let the work speak first.

The Work Speaks Louder Than the Debate

I stopped arguing about AI filmmaking on the internet. Completely. I used to read Reddit threads and Twitter arguments and feel compelled to defend my work, to explain, to justify. I'd spend an hour crafting a response to someone who called me a "talentless hack riding on algorithms." An hour of my life, gone, to change the mind of a stranger who wasn't interested in having their mind changed.

Now I make films instead. the second sample short took ten days from concept to export. Ten days. I made it partly because I had the idea, and partly because I wanted to prove to myself that I could work fast and well without seeking anyone's approval. It's a four-minute dark comedy about a funeral director and his dead mentor's practical jokes. It's funny. It's got heart. Harold Vance's progressively disheveled black suit tells a story all by itself. Nobody watching that film knows or cares which AI generator I used for the establishing shots. They laugh at the whoopee cushion scene. They feel the complicated affection Harold had for his mentor. That's the job.

The best argument for AI filmmaking isn't a comment thread. It's the next film you make, and the one after that, each better than the last, until the quality of the work becomes undeniable even to people who wish it didn't exist. You don't win debates about filmmaking. You win by making films that people remember.

The sample short's Divided Reception

At the third festival where the sample short screened, I got two reactions that I've thought about ever since. The festival programmer — a woman in her thirties who'd watched probably two thousand short films in her career — pulled me aside afterward and said, "I couldn't tell. I mean, I knew from the submission notes, but watching it, I genuinely forgot." That was the best compliment I could have received. Not that I'd fooled her, but that the story had overridden her awareness of the tool.

An hour later, a filmmaker I'd never met approached me at the reception. He'd also made a short film that was screening that night, shot on 16mm film with a crew of eight. "I could tell," he said. "And it bothers me. It bothers me that you call this filmmaking when there was no set, no crew, no one got paid, nobody learned anything on the job. You made this in your house. I don't know what that is, but it's not what I do."

I didn't argue with him. I thanked him for being honest. And I thought about his words for a long time. He wasn't entirely wrong. Nobody got paid on the sample short. No gaffer learned his craft on my set. No production assistant fetched coffee and absorbed the atmosphere of a working film set. Those are real losses, and pretending they don't matter is dishonest. But they're not the whole story. the lead character's grief is real. The audience members who wrote me about their own losses — that's real too. The filmmaking tradition has room for many ways of working. His way is valid. My way is valid. The work decides what matters in the end.

Building Real Confidence

Your first AI film will have problems. Visible ones. Lip sync that drifts. Hands that look wrong. That particular smoothness that marks something as generated rather than filmed. This is fine. Every filmmaker's early work has problems. The question is whether you keep going.

By your third film, you'll have solved half the issues that plagued your first. By your fifth, you'll have developed instincts about what works and what doesn't. By your tenth, you'll be making work that surprises people — including yourself. The improvement curve is steep if you're paying attention and honest about what's not working.

The skills that matter most in AI filmmaking are the same skills that matter in traditional filmmaking. Sound design is still 100% human work. Editing is still 100% human work. Music selection, pacing, performance direction through voice prompts — these are creative choices, not algorithmic outputs. The AI generates raw material. You turn it into cinema. That's not nothing. That's the job. And the people who claim otherwise have never sat with a rough cut at two in the morning trying to find the exact frame where a scene breathes properly.

Pro Tip: Never apologize for your tools in your artist statement. I've read submission statements that begin with caveats about the filmmaker's process, practically begging the reader not to take the work seriously. Don't do this. Your artist statement should talk about what the film is about, what moved you to make it, what you hope the audience feels. If the festival wants technical disclosure, they'll ask for it separately.

Pro Tip: Let your work improve faster than the criticism can keep up. The people who dismissed AI films two years ago were reacting to genuinely rough early work. The tools have gotten better, and so have the filmmakers using them. Make something so good that the tool conversation becomes irrelevant. That's the goal. Quality is the only argument that ever wins permanently.

Pro Tip: Find your community. They're out there — Discord servers, Facebook groups, small festivals that are actively welcoming AI work. Other AI filmmakers are not your competition. They're the only people who understand why you do this, and they will celebrate your wins because they know how hard those wins are. The isolation is the hardest part. Break it.

Common Mistake: Getting defensive in Q&As. I did this after the church basement incident. For my next few screenings, I went in loaded for bear, ready to argue with anyone who raised an objection. I was tense before the film even started. The audience could feel it. Defensiveness makes your work smaller, not bigger. Answer questions honestly, disagree respectfully if you must, and remember that not everyone has to love what you do.

Common Mistake: Trying to hide AI use. It looks guilty because it is guilty. If a festival asks and you don't disclose, you're cheating. If a client asks and you misrepresent your process, you're lying. The truth always comes out, and when it does, your reputation is what suffers. Be honest where honesty is required, and let the work stand on its own where it's not.

Common Mistake: Letting the debate consume your creative energy. I know filmmakers who spend more time arguing online about AI ethics than they spend actually making films. This is a trap. The technology will keep evolving whether you participate in the argument or not. Make your work. Let other people argue about what it means. Your time is finite and the films in your head deserve to exist more than your opinions on Reddit deserve to be read.

What the Future Might Actually Look Like

I think the stigma against AI filmmaking is going to disappear faster than most people expect. Not because everyone will suddenly love it, but because it'll become invisible. Audiences don't care how a film was made if it moves them. They don't ask whether the color grade was done in DaVinci Resolve or Baselight. They don't investigate whether the visual effects were rendered in Maya or Houdini. They watch the film and feel something, or they don't. The tool conversation is always internal to the industry. The audience just wants to be moved.

What if AI filmmaking becomes a recognized category at major festivals? Not a separate, lesser category — just a category, the same way animation is a category, documentary is a category, experimental is a category. We're already seeing festivals add "AI-assisted" or "AI-generated" sections. The next step is normalization. The step after that is no one caring enough to label it anymore. At a certain point, the tools become as invisible as non-linear editing became. Nobody calls a film "Avid-edited" anymore. They just call it a film.

And what if the critics are the ones who end up looking dated? What if, in five years, the filmmakers who refused to engage with these tools are seen the way painters who rejected photography were seen — not as principled purists, but as people who missed a revolution because they were too attached to their familiar methods? I don't know if that's how it'll play out. But I think about it sometimes, late at night, when I'm working on a scene and the tools are doing something that would have been impossible two years ago.

The man in the church basement never came up to me after that Q&A. He left during the next film. But if he ever reads this, I want him to know something: the sample short screened at three festivals, was watched by people who wrote me emails about their own losses, their own static, their own dead wives and husbands and parents they could still hear on old recordings. It moved people. That counts as filmmaking. I'll argue that with anyone, in any basement, at any hour.

The work is the answer. Keep making it.

www.ingramcontent.com/pod-product-compliance
Lightning Source LLC
LaVergne TN
LVHW082333200926
848481LV00012B/3331
* 9 7 9 8 9 9 6 1 8 1 4 6 9 *